Political dynamite. A wake-u
civilization bend rather than
—Roma

The Good Ancestor and *History for Tomorrow*

Analysis, inspiration and sexy 'oven-ready campaigns in this latest fantastic offering from our friends at *Adbusters*.
—Dr Gail Bradbrook, Co-Founder, Extinction Rebellion UK

Capitalism has paved the way to the Extinction of civilization. I am reading the *Manifesto for World Revolution*, because I want to know what is going to happen next.
—Franco "Bifo" Berardi

A deeply honest up-in-your-face-book about our present extremely dangerous world. Shows a way out of the endgame we as humanity and civilization have created for ourselves. The most motivating, compelling and hopeful book I have read for a long time.

Put the *Manifesto for World Revolution* on the reading list for all high school and university students. The book would blow their minds, open their hearts and give them the guts to act like their lives depended on it.

Let me just say it loud and clear: I like everything about the *Manifesto for World Revolution*.

Kalle Lasn and the *Adbusters* team have written a deeply honest up-in-your-face-book about our present extremely dangerous world.
—Úffe Elbæk, Former Minister for Culture of Denmark

Whether they're in advertising, journalism, entertainment, consulting, or politics, people are paid to tell us that life will be OK if we keep on working, shopping, saving and believing. But they're wrong: Gaia has finally gatecrashed the party of industrial consumer society. The *Manifesto for World Revolution* reminds us that the breaking down of dominant systems allows for a rising up of alternative ways to live.
—Prof Jem Bendell, author *Breaking Together*

All human beings are animals of one species whose survival and well-being are utterly dependent on clean air, pure water, rich soil, photosynthesis and biodiversity, which are "sacred gifts" from Nature. The systems we've created to guide and constrain human activity and behaviour, namely law, the economy and politics, pay scant heed to these life-conferring elements and so drive their destruction in the name of human progress.

For decades, leading scientists have been warning that we are on an ecologically destructive path while environmentalists describe a sustainable world but are marginalised as "special interest" groups whose concerns are subordinate to corporate and political priorities.

We have run out of time for incremental change – a new law to rein in corporations, new taxes, financial incentives, new parks, aspirational goals – to get us off the destructive road we are on. We need revolutionary shifts in our priorities and activities immediately. Kalle Lasn provides a blueprint to make that shift.
—David Suzuki

Manifesto for World Revolution

live without dead time

—situ

Also by Kalle Lasn

Culture Jam

Design Anarchy

Meme Wars:
The Creative Destruction
of Neoclassical Economics

MANIFESTO FOR WORLD REVOLUTION

2ND EDITION

Kalle Lasn

with Pedro Inoue
and Bruce Grierson

 Unbreaking

First Edition published 2023 by
Adbusters Media Foundation
www.adbusters.org

This Second Edition first published 2025 by
Unbreaking
An imprint of 5m Books Ltd
Lings, Great Easton
Essex CM6 2HH, UK
Tel: +44 (0)330 1333 580
www.5mbooks.com

Copyright © 2025 Adbusters Media Foundation

The right of Kalle Lasn to be identified as author of this Work has been
asserted by him in accordance with sections 77 and 78 of the Copyright,
Designs and Patents Act 1988.

All rights reserved. No part of this publication may be reproduced, stored in
a retrieval system, or transmitted, in any form or by any means, electronic,
mechanical, photocopying, recording or otherwise, without prior permission
of the copyright holder.

A Catalogue record for this book is available from the British Library.

ISBN 9781917159104 (hbk) / 9781917159111 (pbk)
eISBN 9781917159142 / DOI 10.52517/9781917159142

Printed by Bell & Bain Ltd, Glasgow

Contents

New Ways to Live, Love and Think	1
Birth of a Politics Beyond the Left and Right	9
One Shot One Life	21
Culture Jamming — Let's Have Some Fun	33
The True Cost Revolution	57
Battle for the Soul of Economics	77
Bad Karma	85
Corporate Crackdown!	93
Why Can't We the People Know Everything?	113
A Fundamental Shift in the Nature of Value	125
The Creative Destruction of Neoclassical Economics	133
Occupy Finance!	149
Disillusionment	161
Bioconsciousness	183
Vibe Shift	193
Epilogue	219

for Emi

New ways to live, love and think

One of the great epiphanies of my life happened over thirty-five years ago in my neighborhood supermarket parking lot. I was plugging a coin into a shopping cart when it suddenly occurred to me just what a dope I was. Here I was putting in my quarter for the privilege of spending money in a store I come to every week but hate, a sterile chain store that rarely offers any locally grown produce and always makes me wait in line to pay. And when I am done shopping, I'd have to take this cart back to the exact spot their efficiency experts have decreed, slide it back in with all the other carts, rehook it, and press a button to get my damn quarter back.

A little internal fuse blew. I stopped moving. I glanced around to make sure no one was watching. Then I reached for that big bent coin I'd been carrying around in my pocket and I rammed it as hard as I could into the coin slot. And then with the lucky Buddha charm on my keyring I banged that coin in tight until it jammed. I didn't stop to analyze whether this was ethical or not — I just let my anger flow. And then I walked away from the supermarket and headed for the little fruit and vegetable store down the road. I felt more alive than I had in months.

Much later I realized that I had stumbled on one of the great secrets of modern urban existence. Honor your instincts. Let your anger out. When it wells up suddenly from deep in your gut, don't suppress it — channel it, trust it, use it. Don't be so unthinkingly civil all the time. Don't let small injustices pass. When the system is grinding you down, unplug the grinding wheel.

Once you start thinking and acting this way, once you realize that consumer, corpo-capitalist system we live in is by its very nature unethical, and therefore it's *not* unethical to fight back; once you understand that civil disobedience has a long and honorable history that goes back to Ghandi, and Thoreau and King, Jr.; once you start trusting yourself and relating to the world as an empowered human being instead of a hapless consumer drone, something remarkable happens. Your cynicism dissolves. Your instincts sharpen.

Direct action is a proclamation of personal independence. You act. You thrust yourself forward and intervene. And then you hang loose and deal with whatever comes. In that moment of decision, in that leap into the unknown, you come to life. Your interior world is suddenly vivid. You're like a cat on the prowl: alive, alert and still a little wild.

We're raised to be polite and civil, but sometimes polite is exactly the wrong thing to be. Polite lets bullies off scot free. Polite lets corporations walk all over us. Polite lets our leaders do nothing about climate change. Polite lets cancerous systems grow until there's no stopping them.

That's pretty much where we are at now.

To those who doubt that massive global breakdown is coming I say: *It has already started.* Look around. Everything we've considered stable and enduring for centuries is slipping away. Corporations rule by fiat. Big Finance plays countries off against each other. Economists push for more growth while ecosystems crash. Algorithms swing elections. Unanswerable lies proliferate. And lately, brutal crazy stuff: The rise of ruthless evildoers like MBS, Sisi, Putin, Lukashenko, Netanyahu and Xi. Rape, torture and starvation are now routinely used as weapons of war.

As temperatures rise, natural systems collapse and social structures crumble, there'll be water shortages and food riots. The

Revolution is a force that gives life meaning

number of failed states will jump from the current twenty to thirty, then fifty-plus. Refugees by the millions will first beg and then fight their way into every corner of the rich world.

And then, in a geopolitical moment of reckoning, a conflict will escalate out of control. Rage will ignite even in the most tranquil minds. All hell will break loose. Nobody will be able to control it. Hundreds of millions, maybe billions, will perish in a massive die-off. Those who survive will huddle up and cower through a dark age that could last a thousand years or more.

This is not hyperbole. The fate of this six-million-year experiment of ours on Planet Earth is hanging in the balance, if only we had the clarity of mind to see it.

Is there any hope left? Can we wake up from this nightmare we're living in?

At *Adbusters,* we often sit around blue-skying. We debate what kinds of wild interventions might actually work, what leaps of imagination could, if adopted at scale, nudge this doomsday machine of ours back onto a sane, sustainable path.

Maybe we can get a corporate charter revocation movement going and wipe ExxonMobil off the face of the earth.

Maybe we can conduct small acts of subterfuge — like placing *OUT OF ORDER* signs on ATMs — that inspire a multinational grassroots campaign that brings Big Finance to its knees.

Maybe we can design a new global marketplace in which the price of every product tells the ecological truth.

The poster that helped spark Occupy Wall Street

Maybe we can kindle a paradigm shift in the science of economics — one that measures progress differently and understands that growth cannot go on forever.

Maybe we can launch a *Mental Liberation Front*, kill off the ad-infested platforms and revive the dream of the FREENET.

Maybe the artists, designers and architects of the world can set in motion a vibe shift that inhibits impulse, scrambles habits and modulates desire.

Maybe we can spark another global big-bang moment like Occupy Wall Street. And maybe this time it will gain so much momentum that nothing and no one can shut it down.

This book is a field guide to a new world order. We're calling for a people's revolution — a new way to live, love and play on this planet.

Birth of a politics beyond the Left and Right

The English revolution, the American revolution, the French, Russian and Chinese revolutions — insurrections come and go. It's how we progress. It's how societies leap into new orbits.

Up till now, political uprisings have always been local. But today we are witnessing the birth of a new kind of rebellion, one that operates completely outside of geographic borders and political structures.

A Third Force.

All over the world, people are coming together online to craft new ways of influencing policy, wielding power and self-governing ourselves. We're moving beyond any domestic political paradigm and for the first time in human history, starting to think and act as a global community.

We are the advance guard — folks who yesterday were quite comfortable in their jobs, their families, their communities, only to be lurched awake to discover that they are victims of a corpo-capitalist Ponzi scheme beyond imagining.

We're not a political party. Nobody voted for us. We don't identify as Left *or* Right.

Like any real breakthrough political force must be, we are a synthesis of opposites. From now on, politics will no longer be the usual slugfest between ideological shadow-selves. A third player with a radical new agenda has joined the fray.

From now on, politics will look like this:

The Third Force

Our elected governments, the United Nations, the World Bank and the International Monetary Fund will continue to function much as they always have. They will repair the roads, collect the taxes, run the courts, deliver the mail, give financial aid to developing countries and send peacekeepers to conflict areas. But now they'll have to deal with this unsilenceable new voice demanding systemic transformation on multiple fronts.

When We the People feel our leaders aren't paying attention, are mishandling things and veering off course; when the yawning gap between the haves and the have-nots grows too wide; when corporations become too arrogant, Wall Street too greedy, secrecy too pervasive and surveillance too invasive; when the norms, values and precepts that underlie our way of life are violated . . . that's when We the People will rise up and set things right.

Our power is in our numbers. Get enough people aligned and their shouts become a kind of flocking signal.

The Third Force is a new evolutionary cluster, a group that forms when historical conditions are right, *because it has to*, to accelerate change and give humanity a chance.

The difference this time is that the insurgents have an immeasurably potent new bit of technological leverage at their disposal.

The Internet has reversed a centuries-old power dynamic. Social media is now a third arm of democracy, along with the law and government checks-and-balances. It allows people to share what they're dreaming up in real time, and get behind the best of it. Millions can converge on hundreds of activist websites like abillionpeople.org. That's enough to launch massive boycotts that bring wayward corporations to their knees. Enough to organize global big-bang moments when people swarm their cities demanding systemic change. Enough to stop wars.

We the People now have the ability to poke our noses into every policy debate, every election, every UN Security Council decision, every military conflict.

And we can penetrate deeper still, right into the beating heart of our global system, shifting paradigms, reversing money flows, triggering reformations.

The Third Force will burn its initials into the culture with the focused heat of the holy fire you're holding in the palm of your hand.

The Net is what we make of it. It *can* be a quagmire where you're sucked in, submit your attention to someone else's agenda and come out feeling diminished. But we've also seen flashes of what the Net can do when it's wielded as a sword of resistance. How quickly it can whip up a social transformation: The Arab Spring, #OccupyWallStreet, #MeToo, #BLM, #FreePalestine.

We're just learning how to harness the power of connectivity. Not in a superficial way, reacting every few minutes to some new insult, but slowly learning to be *agents* of transformational change, burrowing down to first principles to prevent chaos.

If we play our cards right, maintain focus, all the while keeping it loose, keeping it nimble, improvising riffs, being systems thinkers of the highest order, then we can harness humanity's collective power and become the dominant political force of the 21st century.

THE METAMEME INSURRECTION

Donella Meadows, the environmental scientist who co-authored *The Limits of Growth*, urged us to burrow below the surface. In her iceberg model of systems thinking, the events that make the news are only a small part of what's going on.

One level down lie trends. Repeating patterns. Beneath that are structures, systems, paradigms, forms.

The deeper you go, the more leverage you have. It takes pressure applied at the bottom — the level of our most ingrained values, norms, assumptions and beliefs — to bring about meaningful, lasting, systemic change.

We are now waking up to the fact that the world runs on a handful of well-entrenched, largely unquestioned precepts. Most of us take it for granted that economists know what they are doing. That toxic financial instruments like derivatives and credit default swaps are business as usual. That flash trading is an efficient way to run stock exchanges. That money can move freely across borders but people can't. That advertising is harmless. That secrecy is a normal part of democracy. That arms trading cannot be stopped. That no matter how heinous a crime a corporation commits, it's untouchable.

But now, caught in an existential crisis with no obvious way out, we begin to question these hidden coordinates of our reality and start thinking about a new operating system for Planet Earth. We hatch a new grand narrative, a set of ideas so fundamental, so systemic, so profound that a sane sustainable future is unthinkable without them.

And then we deploy them.

We crack the global mind on seven critical fronts:

On the ecological front, we get Adam Smith's invisible hand working for us instead of against us. We lay a plan for a capitalist reformation — a painful but necessary move towards a new kind of global marketplace, a *bioeconomy* — in which the price of every product tells the ecological truth.

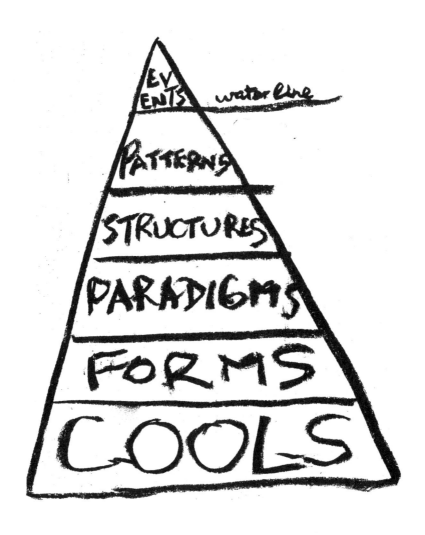

On the corporate front, we launch the mother of all boycotts against one of the most criminal corporations in the world and wipe it out. Once the mightiest has fallen, the rest will follow, and a *Corporate Charter Revocation Movement* will be born.

On the financial front, we force our leaders to eliminate all tax havens. We march worldwide for a 1% Robin Hood Tax on all stock market and currency transactions. We push for new rules to cool down insanely feverish bot-run trading. We propose the simple idea that you must hold a stock for 24 hours after its purchase before selling it.

On the economic front, we set up subversive cells in the economics departments of universities around the world, and start disrupting classes, popping posters in the corridors and nailing manifestos on professors' doors. We expose the current science of economics as a *hindering* profession, disastrously out of tune with planetary biorhythms. We seed a revolutionary leap in economic thinking — a paradigm shift toward a new *bionomic* model that works in the real world.

On the political front, we push to rewrite the constitutions of nations, recalibrating the role of humans in the ecological chain. We make secrecy taboo in all but the most sensitive areas of national security. We clamp down on the arms trade. In nation after nation, we propose a constitutional amendment for global adoption, calling for national referendums requiring 50 percent of voters to assent to any war.

On the psychological front, we expand our concept of human rights beyond our physical bodies into the digital realm; your data becomes part of your new hybrid self. Then we launch the *Mental Liberation Front* (MLF) and start reclaiming our mental space — hacking into the networks and monkeywrenching the algorithms that are manipulating us and undermining our individual and collective will.

On the aesthetic front, we call on architects, artists, designers and creatives to transform the ambient tone of the world. The way it feels to walk around our cities; the mood of watching television; the knack and smack of navigating the internet; the emotional valence of money and status; the way it feels to be alive today. We step off the godless and immoral straight line we've been stuck on for the past century and learn how to wobble again.

We are a new norm-setting force. With these metamemetic transformations as our playbook, we can jolt this experiment of ours on Planet Earth back onto a sane, sustainable path.

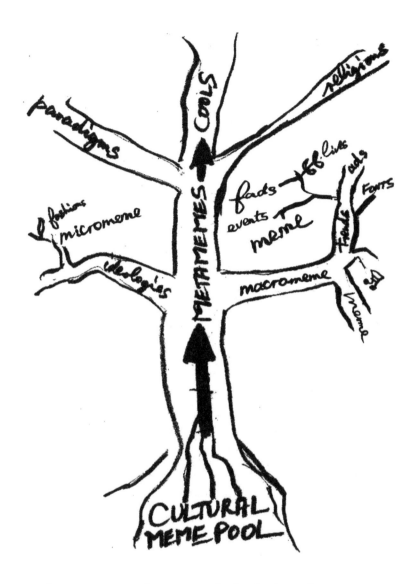

At this point some of you are no doubt thinking this is crazy, impossible stuff. Monumental social, political and cultural heaves like this take generations, sometimes centuries, to kick in, if they ever do. I hear all the time from critics about pretty much every one of these metamemes: "It'll never fly."

And I always respond: "It'll never fly *in your world.*" The business-as-usual world. But that world will soon be gone.

The only thing that can save us now is a massive change in the way we think about living on this rock — a quick and mighty swerve toward common goals and shared resources and long horizons.

As our ecosystems crash, as our minds become muddled, as the center gives way and we start spiraling into a new dark age, millions of us will embark on a revolutionary adventure of social transformation. Our mission: to demolish corpo-consumer-capitalism as we know it.

Isn't it ironic? Once we realize we're doomed if we do nothing, that puts us in an oddly liberating place. It means we have nothing to lose. Anything and everything suddenly becomes possible, and all the homilies, banalities, and precepts we've taken for granted for centuries begin to crumble.

We're in a geopolitical power-shift moment. The hierarchical, top-down power structures that have ruled the world for thousands of years are collapsing before our eyes. The street now holds unprecedented sway. We know what we want. After centuries of rule by kings, emperors, tyrants, mad men, fascists, communists, military dictatorships and mega-corporations, We the People of the world are now ready to take charge of our own destiny and start calling the shots from below.

And if the leaders, politicians, intellectuals and pundits of the old-world order refuse to listen to us, then, armed with our metamemes, we rush into the streets and get the job done by the sheer power of enraged swarms of humanity howling for deliverance from a future that does not compute.

We are the Third Force.

We will spark the first ever global revolution and win the planetary endgame.

I'm about to give you the codes.

Are you ready?

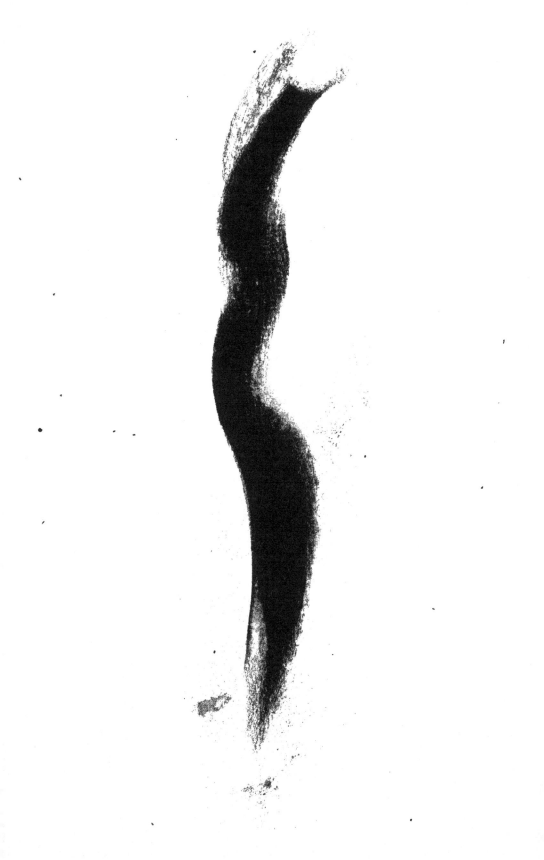

One Shot One Life

Estonia, January 1943: The Red Army busts through the Leningrad Blockade and marches into the capital city of Tallinn. Huddled in the dark in a root cellar a hundred miles to the West, I can smell the raw, dank earth. Just me and my mother, father, and sister, surrounded by sacks of potatoes. Above, sirens blare. The Russians are hardly liberators. They're almost worse than the Nazis. They will soon cull the population of Estonia to a million people, and who knows what fate befalls my family if we don't manage to flee on one of the last boats out.

We end up in a displaced-person camp — the first of many. Food is scarce. Some weekends we walk deep into the German countryside to barter with farmers. On one occasion, my mother exchanges her precious fur coat for a bag of eggs and vegetables.

My father lands a part-time job at the UN relief organization UNRRA — and to the great joy of the whole family he sometimes brings home a can of Spam, baked beans, a jar of strawberry jam.

As the repatriation efforts begin, we're told to gather our few belongings yet again. One day a tall American soldier with an aw-shucks smile scoops me up and presses a Hershey bar into my hand.

The next five years are a blur of barracks and cots and lineups and boredom. We are in and out of refugee camps until I am seven years old. Then come another five years of refugee camps in Australia.

My whole childhood unfolds in a bubble.

We've been kicked out of our own country by the Communists, leaving everything behind. So anything that smells of collectivism is anathema to my father.

I can hear him muttering late at night: "Those Commie bastards. . ." I pretend to sleep but listen intently to the drunken banter of Estonian expats . . . "I'd like to line those fucking commies up against a wall and mow them down one by one . . ."

But I can also hear them argue, other nights, about a book called *The Decline of the West*. The idea that Western civilization is decaying and may soon die off.

My father was a tennis champion. His actual job was as a lawyer for the Estonian government, but who he *was* was an athlete — and something of a national hero. Before WWII started, he was the number one singles tennis player in the Baltics. He captained Estonia's Davis Cup team and played Wimbledon. But he could never beat Sweden's Kalle Schröder. That's why I think he named me after him.

When I stepped onto campus at the University of Adelaide, I'm not sure I had a political bone in my body.

I noticed two types of students running around. There were the beer-drinking righties with their Bertrand Russell, A.J. Ayer and Gilbert Ryle, more British than the British. And then there were the Gauloise-smoking, coffee-sipping Lefties, with their Kierkegaard, their Nietzsche, their Sartre and Camus. I wondered: Which side am I on?

I studied mathematics because my parents, with their refugee mentality, pushed me into the sciences so I could make a "decent" living. My heart, though, was set on philosophy.

The philosophy department in Adelaide was firmly in the British "logical positivist" camp. Every word, every sentence, every thought had to be rigorously scrutinized for logical purity — all religious and metaphysical nuances were weeded out by Occam's Razor.

Wading through Ludwig Wittgenstein's *Tractatus* I found a few gems: "What we cannot speak about we must pass over in silence." In his Cambridge lectures he would often bury his head in his hands and mutter, "I am a fool … I am a fool." At least that was the legend we'd been told.

Later I read his *Philosophical Investigations*. Here the hard dogmatism of his *Tractatus* gave way to something softer, more forgiving. A few of his aphorisms had a mystical ring to them. Wittgenstein fought for Austria in WWI. After glimpsing the Russian enemy for the first time, he wrote in his diary, "Now I have the chance to be a decent human being, for I'm standing eye to eye with death."

The punch of that quote stayed with me. Maybe because it carried a message that something more than logic must be at stake. Wittgenstein's last words on his deathbed were "Tell them I've had a wonderful life."

A friend introduced me to classical music. His love for Berlioz, Bach and Beethoven bordered on the obsessive, and some of it rubbed off on me. I joined a record club, and once a month when the disc arrived, I'd put my head close to the gramophone late at night. The dark tones of Sibelius. The sublime quiet movement of *Beethoven's 9th*. The triumphant cannon blasts in Tchaikovsky's *1812 Overture*. And later in life, the dissonant forebodings of Bruckner and Mahler.

In my final year, someone passed me a copy of *The Outsider* by Colin Wilson, one of Britain's so-called "angry young men."

Whoa.

Here was a kind of grand, unifying theory of the fully alive Western man — a spiritual and cognitive seeker, a freelance warrior soul. You could learn a lot, Wilson was saying, from the intrepid scouts who had gone to the edge and reported back. Kafka. Nietzsche. Camus. T.E. Lawrence. Dostoyevsky. Nijinski. Van Gogh. What these artists were chasing was intensity — a rare thing in a culture that seemed frightened of such animal impulses. All of them had experienced what Wilson called "moments of vision." Some of them were a little nuts. But they saw further, vibrated faster, sucked the stars out of the sky.

The Outsider was written like the author thought he was going to be shot at dawn. I ripped through it. It was ecstatic. It promised deliverance to a higher plane of being. The idea was that you could catch lightning in a bottle *if* you just had the guts to stand out there in the thunderstorm, with an open heart.

Sartre thought everything is pure chance — "life is a useless passion"— but my feeling has always been the opposite: that we humans are on the verge of an evolutionary leap to a higher plane.

Then one day I discovered what Wilson's ecstatic vision of rebellious authenticity really looked like.

It was late one Saturday night. After a long bull session in somebody's room, drinking beer, getting high on something, a few of us ended up in a loft in the warehouse district. The tenant was a rawboned guy with long messy hair and a beatnik beard. On his wall, a smash of posters, newspaper tear-outs, scribbled musings. Behind his bed a stolen *DETOUR* sign.

This guy's whole existence said: You can rebel.

The Australian government paid my way through university, and in exchange I gave them three years of my life. I played computer war games for the Department of Defense, helping the government decide what kind of military hardware to buy. As soon as my contract was up, I hopped on a boat heading for Amsterdam.

But the boat made a stop in Yokohama. There, unfolding all around were the mysteries of a culture that turned everything I believed inside out. I was entranced. Three days later the boat left for Amsterdam. I wasn't on it.

I got a job with a consulting company, and soon after started my own marketing firm, grandiosely called International Computer Research (ICR), in Tokyo's Roppongi district. I made tons of money. That's no brag: in the booming Sixties anybody could.

I hit the clubs most nights with my ad-agency pals. Behind the scenes, I dabbled in judo & Zen. Savored Kawabata's *The Sound of The Mountain*, Yukio Mishima's *Forbidden Colors*, Osamu Dazai's *No Longer Human*. Then one day... Basho.

> a quiet temple pond
> a frog jumps in
> plop

What made Japan so fascinating was that that this sparse, ancient Buddhist aesthetic could survive the blast of go-go postwar capitalism. The legendary calligrapher Sesshu would spend many days preparing himself — and then put the whole weight of his soul into one fierce stroke of the brush.

> one shot,
> one life

One afternoon I caught Akira Kurosawa's *Ikiru* and walked out of the theatre stunned. Then I discovered Yasujiro Ozu. On the surface not much happens in Ozu's films. In *Tokyo Story* there's a scene of an old couple, who have come from the country, sitting on a park bench in the sunlight. They don't speak. Nothing happens. The scene lingers for what feels like forever. It is incredibly profound.

In the bars, night after night, I met disillusioned American GIs on R & R leave from the Vietnam War. They told me harrowing stories of their lives over there. Many of them went AWOL, hiding in their girlfriends' apartments until the military police tracked them down and dragged them back to their units.

Then I got wind of another uprising. It had started in Paris' Latin Quarter and quickly spread around the world.

The English-language *Japan Times* reported that furious young men and women were rising up against consumer capitalism. Students, artists, nurses, doctors, bus drivers. They occupied campuses and factories and hospitals, singing songs, issuing manifestos. All over Paris they sprayed slogans like *Live Without Dead Time* and *Under the cobblestones, the beach!*

Through the speakers of GI radio came a siren call out of Haight-Ashbury. Otis Redding had a pretty good view of things from the dock of the bay. "If you're going to San Francisco, wear some flowers in your hair..." That lifted my spirits every time I heard it. The spirit of '68.

It was time to see what it was all about.

"Feed your head," Jefferson Airplane sang.

Grace Slick would later say that song ("White Rabbit") was inspired by Miles Davis's *Sketches in Spain*. She said it meant follow your curiosity. But everyone knew what it really meant.

LSD was the original love drug — ecstasy before there was Ecstasy. An empathy machine. It dissolved boundaries between individual people and between people and the world. It made us all truly *get* the idea of the collective. No one understood "we're all in this together" like the hippies.

The people I was meeting in San Francisco were like the guy in that smoky loft in Adelaide.

There are moments in history "when millions of people surge into the streets and refuse to leave until real shifts are made in the social order. These moments accomplish in days what takes history years or even decades."

The American historian George Katsiaficas called it The Eros Effect.

In a blink there is a total values shift. "Instead of patriotism, hierarchy or competition being dominant, people construct new values of solidarity, humanity, of love for each other."

Mainstream sociologists consider these cultural heaves a kind of mass hysteria, bursts of madness. And yet, as Katsiaficas put it, "when we look at them from the bottom, from the perspective of ordinary people, these are moments of freedom. Leaders are unable to control the love of people for each other."

In these moments, something universal within us is ignited. And radicalization and revolt follow as day follows night.

In San Francisco's Haight-Ashbury district, I found myself in the middle of an oxymoron — a community of mavericks. We felt a lot of things were broken, and we had ideas about how to fix them. We'd take this old world our stuffy parents believed in, and set it free.

I bought a used VW beetle and drove all over America.

I was 200 miles from Memphis when Martin Luther King was shot and the riots broke out everywhere at once. I headed south to Mexico, then deeper into Central America. Through Guatemala and Nicaragua (when Anastasio Somoza was still in power). Through Costa Rica and Panama. I spent half a year in what was then still called British Honduras.

Chatting with all the other five-dollars-a-day backpackers, I learned the real story of how the Panama Canal was built. How, before Castro, Cuba was America's gambling brothel. How the CIA trained death squads in Brazil and propped up Pinochet in Chile and let Baby Doc Duvalier run amok in Haiti.

The same questions kept bubbling up: What *is* America? The country was built on slavery, their Monroe Doctrine has been doing dirty deeds in Central and South America for a hundred years. They have a long history of interfering. Just as they were doing again now in Vietnam.

Yes, America saved the world from the Nazis in World War II; but — at least at the level of their administration — they are not the beautiful people on the right side of history that I had so fervently believed in. It was like handing over my American History textbook and upgrading to Howard Zinn.

But on good days you could ride the cultural riptide out beyond those shoals and escape. The Beatles. *The Graduate*. Ginsberg. "Follow your inner moonlight / don't hide the madness."

Wherever I went I'd find a rep cinema and slide into the anonymous dark to catch a matinée.

I returned to Japan, married my soulmate Masako, and together we decided to start a new life in Canada.

The National Film Board of Canada had a reputation as one of the most exciting documentary filmmaking institutions in the world. No artist could fail to feel its pull. As soon as we arrived in Vancouver, I bought a 16mm projector and all I did the first few months was watch NFB films over and over again.

It blew my mind.

Norman McLaren was a gateway drug to his brilliant, troubled protégé Ryan Larkin, whose psychedelic short film *Walking* left me in an existential reverie. That you could take the private world in your head, your wildest and most profound stirrings, and splash them across a moving wall and still call it truth was a revelation.

Light, hand-held cameras and portable sound recorders were opening up new ways to make documentaries. All kinds of innovative cinéma vérité moves were suddenly possible.

A bunch of us started a film commune in the house on False Creek that's now the site of the *Adbusters* offices. We made experimental films and subsidized ourselves by selling mandarin oranges from roadside stands at Christmastime.

The best thing we did was *Schizophrenic Superman* — a collage of comic book cutouts assembled to the beat of a square dance record I found in the basement. It was a big hit in the weekly film showings we held on Sunday nights, complete with cheap homemade beer infamous for giving everyone the trots.

For the next 15 years I made films. My first real success was *Ritual*, a half-hour documentary about Japan shot mostly with a hand-held Bolex camera. PBS picked it up and it aired repeatedly across their affiliates. This was my passport into the NFB, and over the next 10 years I made a series of documentaries for them about Japan and the global economy.

In the spring of 1989, British Columbia's forest industry launched a slick multimillion-dollar campaign called "Forests Forever." On billboards, in newspapers and on TV, the industry bragged about the marvelous job it was doing managing the province's forests. Rest easy British Columbia, their ads proclaimed, you've got nothing to worry about, you've got "Forests Forever."

This made a few of us green-minded filmmakers very angry. We decided to make our own 30-second TV spot telling the other side of the story: that British Columbia's old-growth forests were being decimated at an alarming rate and the future of this precious resource was far from secure.

But when we tried to buy airtime, we were in for a shock. None of the TV networks would take our money.

We decided to fight back. We fired off press releases, got our story into the newspapers, started a newsletter, launched a legal action against the Canadian Broadcasting Corporation (CBC).

We never did get our counter-ad on the air, but in a victory of sorts, the CBC finally relented and stopped running the *Forests Forever* campaign. We were ecstatic.

Meanwhile, our newsletter had become quite popular. It was exhilarating to have a voice, to speak truth to power. When I told my journalist friends I was thinking of starting a magazine, they warned me against it. ("Your garage will fill up with unsold copies. You'll go bankrupt. Your wife will divorce you.")

Taking a cue from a little NGO startup across town named Greenpeace, we called ourselves the "Journal of the Mental Environment."

In the beginning *Adbusters* was a typical Lefty rag, full of impassioned essays, rants against evil corporations and fervent calls for radical change, with nary a photograph or cartoon for comic relief. We mimicked the *Utne Reader*, *The Nation*, *Mother Jones* and championed all the Lefty causes of the time.

We decided to rely only on subscriptions and newsstand sales. We vowed never to sell space to advertisers. And bit by bit we embarked on an aesthetic journey. A journey, you might say, to get off the grid — to abandon that boring, text-centric look of most progressive magazines . . . to create a magazine that was less about the content you "consume" than a river you jump into.

After a few years, *Adbusters* was all over Canada. And then the US. By the mid 1990s we were selling briskly on newsstands around the world.

What everyone remembers from that period are the spoof ads on our back and inside covers. Absolut Vodka. McDonald's, Philip Morris, Obsession, Nike. We went after any mega-corporate ad campaign that pissed us off. These companies had spent millions to build a nuclear glow around their brands, and now, in deft, judo-like moves, we threw them onto the mat with the power of their own momentum.

Culture Jamming — Let's Have Some Fun

Frederick Hunterwasser famously said, "*The straight line is godless and immoral,*" and at *Adbusters* we've been coming back to that sentiment again and again because it seems to contain an aesthetic secret.

The logic freaks have had it their way for a very long time and all they have to offer is more of the same: more technology, more rationality, more consumption, more surveillance, more control.

But now suddenly, up from the greasy boiler room of the heart comes . . . a *toneshift*! Hyperrationality rolls over and a touch of divine insanity creeps in: *To hell with straight line thinking — let's learn to wobble again!*

This was the conclusion we came to after we'd put a couple of issues of *Adbusters* in the can.

We saw conventional newsstand magazines as poisoned mindspace, because the ads rule. They slam down and crush whatever little tender shoots of thought and feeling you're trying to grow. So our decision to never run ads, you could say, was our first aesthetic choice.

The goal was to get off the grid; to create a magazine that was less about the content you "consume" than a river you jump into and are swept downstream . . . to make everything *flow* . . . to create a mind journey that moved like a movie.

Yasujiru Ozu measured the worth of his films by how many bottles of sake he and his collaborator drank over many weeks creating the script. Every issue of *Adbusters* was a little like that. You never knew what you were going to get. That's because *we* never knew what we were going to give you — until the deadline hit.

Breaking rules is deliciously addictive. We reveled in jamming the traditional norms of print.

We killed the page numbers, because they are just speed bumps that break the spell . . . like someone pulling out a measuring tape in the middle of sex.

And who needs a table of contents? Or article headings? Or even a contributors page? On a fiery mind journey, who needs to be interrupted with pronouncements?

Then we took the letters to the editor and sprinkled them throughout (a very democratic move).

Then we ripped ads we hated out of other magazines and used them as counterpoint (a neat reversal of capitalist appropriation).

We started dropping in cartoons and poems, not randomly like *The New Yorker* does it, but carefully, where they'd resonate with flotsam in the flow.

And we mucked with punctuation and grammar. (Why be so anal with language?)

Every which way we could, we abandoned the sanitized, soul-destroying modernist look.

What was this thing, tattooed with doodles and coffee spills, jammed with random poems and newspaper headlines snipped with sewing scissors? How do you even read it? From front to back or from both ends at once, clashing triumphantly in the middle? It was more like a movie than a book. We'd grab our readers by the throat on the cover and never let go. We'd create a jump-cut in the human imagination. You pick up the magazine and out falls a black spot — an anti-logo representing people power. Activists started putting them up on walls. In the cities they replicated like spores, or mold, nature reclaiming human minds.

When one person gallops through town breaking rules, they're an outlaw. When a whole community starts thinking like outlaws, now what you have on your hands is a seismic cultural event.

Imagine this: it's a prime-time evening in America, circa 1995, and three million people are watching their favorite show — maybe it's *Cheers* or *Murphy Brown*. At station break, a batch of cosmetics and fashion ads come on featuring perfectly beautiful models peacocking down the runway. Then … *POW!* Cut to a ribby, bulimic Kate Moss-lookalike bent over the toilet puking her guts out. Our fifteen-second mindbomb ends with an outshot like the voice of God:

"The Beauty Industry is the Beast."

What the fuck?!

America, you've been slimed.

That's culture jamming, baby!

Ridiculous, yes. But is it any more ridiculous than what we were trying to disrupt? The passive acceptance of a one-way flow of marketing messages that tickle your id and sculpt your feelings and shape your values and dictate what you should do with your time and your energy, your politics and your dreams?

What I felt back then, a quarter-century ago, is what I still feel today: that one-way corporate mindfuck is approaching a level where a massive pushback is inevitable. An anti-corporate uprising will be the next great social movement of our time. The fact that three other major social movements leapfrogged it makes me more, not less, certain of its inevitability. The beauty and power of #OccupyWallStreet, #MeToo, and #BlackLivesMatter is that they're training people to detect hidden patterns — to notice who's being shafted. We don't see logos, we don't hear corporatese, we don't feel overly angry about their total surveillance, or too let down when heinous criminals like Wells Fargo, Purdue and ExxonMobil walk free. We're so far up the butt of America Inc. that we're on the side of injustice without even realizing it.

One of Adbusters' first culture jams, circa 1993, 42nd St., Manhattan, NY.

This was the work, then, back in the early days of *Adbusters*: to smash the commercial monopoly on the production of meaning. The first generation raised on television was mounting a resistance, pulling together a game plan on the fly.

We were inspired by how a handful of low-budget anti-smoking campaigns had toppled the billion-dollar PR might of Big Tobacco. Suddenly, on every cigarette pack, sickening pictures of diseased lungs. On TV, celebrity actor Yul Brynner, just before he died of lung cancer, looked you straight in the eye and said: *"Whatever you do, don't smoke!"* These jams basically rewired our response to images of smoking — from seduction to revulsion. Now every time the Marlboro cowboy lit up, people winced. Philip Morris realized that every dime they spent on advertising was actually doing self-damage. So they stopped.

We were chuffed. A well-crafted, low-budget attack was all it took to bring these fuckers down. It isn't about the size of your war chest; it's about the condensed power of your message. Small and cunning beats rich and lumbering and on the wrong side of history. Our upstart "journal of the mental environment" had a chance. With deft judo-like moves we would throw mega-corporations to the mat with the power of their own misplaced momentum. We'd beat them at their own game.

If it could work on Big Tobacco — one of the biggest, baddest industries of all — we reckoned it could work across them all — food, fashion, automobiles, communications, finance. Every area of our lives where someone had broken in and shoved their swagger down our throats.

We decided to lead the charge. We'd detourn their logos, dick around with their ad campaigns, inflict brand damage and duke it out with them on TV. We'd neutralize the cultural power these

Anti-ad in Adbusters' inaugural issue, summer 1989.

corporations had claimed as their own. McDonald's, Absolut, Calvin Klein, Nike. None were too big to fall.

I was in full double-agent mode now. Every advertising trick I learned in Tokyo could be used in reverse. Instead of showcasing the client's unique strength, we'd identify their unique vulnerability. We'd make folks understand — not intellectually but *viscerally* — what is fraudulent about their seductions.

Take Absolut Vodka. Their flood-the-zone PR strategy had them on the back covers of magazines on newsstands around the world. What's the Achilles heel of Big Spirits? Same as Big Tobacco. It's showing people what *really* happens when you use their product as directed. What's alcohol actually about? Not the shining promise of turning you into a hilarious party animal who gets laid in the car, but brain-fog, impotence and puking on your shoes. Career suicide. Actual suicide. AA meetings. Death and suffering and humiliation: Make folks feel it.

A mindbomb delivered in a goodie bag. What was it Billy Wilder said? If you're gonna tell the truth, you'd better be funny or they'll kill you.

Absolut had a global print campaign centered on the shape of their bottles, and we countered with a riff on the same theme that got a lot of attention. It amused everyone but Absolut. They sent an international law firm with a cease-and-desist order.

This was actually pretty intimidating. They implied they could destroy us if we didn't stop our spoofing campaign and of course they were right. Fear works on a power differential, and you never know how far they're willing to go.

Back cover of Adbusters issue #12, Summer 1995

But then we fought back and the press got hold of the story. They picked up on the David-vs-Goliath angle. Suddenly the dynamic reversed. Absolut looked like a bully and they quickly backed off.

It was a delicious moment. Absolut had scared the hell out of us. But it became clear we had also scared the hell out of them. They knew we had the people behind us and could inflict severe brand damage.

When Absolut first came after us, we added an "e" to the word on our spoofs for legal protection; but when they pulled their lawsuit we exhaled, put the five-finger flag to our nose and took the "e" out again.

We could tell Nike CEO Phil Knight was rattled too when he got wind of a billboard spoof we were planning near the company's

HQ in Beaverton, Ore. Phil could not countenance the thought of an unswooshing message welcoming his employees as they drove to work in the morning.

So now here we were again, faced with another multinational bringing all its horses. But this time we knew better than to be intimidated. When reporters called us up, one minute we're talking about Tiger Wood's swooshy smile, and the next minute we're talking about Indonesian sweatshop labor. The stories come out, Phil Knight looks like an asshole and his swoosh has lost a bit of its nuclear glow.

And I realized something. The wild, sledgehammer-on-a-fly overreaction of these multi-billion-dollar corporations exposes how fragile is this thing called cool. And how you can't impose it on people. Real cool is a revolutionary impulse — a reaction to power. Top-down corporate cool is a fraudulent algorithm, and deep down the big guys know it. That's why they're scared. The emperor is buck naked. We the People rule!

Buy Nothing Day was our first big success. The fledgling environmental movement was banging the drum that overconsumption was at the core of all our planetary woes, and we caught the zeitgeist. We produced a 30-second TV spot featuring a burping pig wallowing on a map of North America. The voice-over

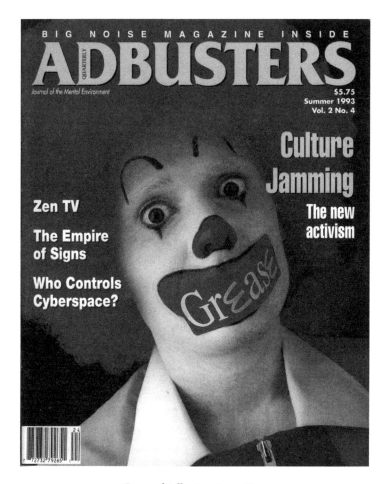

Cover of Adbusters issue #8

threw out some sobering figures about consumption levels and chided that *"the world could die because of the way we North Americans live. Give it a rest, November 26 is Buy Nothing Day."*

The meme spread. Buy Nothing Days popped up in Vancouver, Seattle, San Francisco, Austin, New York. It jumped the ocean, to Sydney and Melbourne, then to London (where they called it No Shop Day). Within a few years, *Buy Nothing Day* was a yearly celebration of frugal living in more than 60 countries.

We learned that if you let loose a tantalizing meme at just the right cultural moment, it can hit the public imagination with incredible force.

We then produced a flurry of 30-second TV spots. First we called them "anti-ads," then "uncommercials," then "subvertisements," and finally "mindbombs."

There was *Autosaurus*, a takedown of the auto industry featuring a rampaging dinosaur made of scrapped cars and ending with a vision of the future: folks riding bicycles over the car-free plain.

There was *Obsession Fetish*, a critique of the fashion industry featuring a bulimic Kate Moss lookalike.

There was a whole series of *Tubehead* spots which spoofed TV addiction and hatched with a slogan that's only become more relevant in the age of mass surveillance: "The Product Is You."

Bull In the China Shop mocked economists for mis-measuring progress. When everything — even ecological disasters — makes the Gross Domestic Product (GDP) go up, something doesn't add up. "Economists," we suggested, "must learn to subtract."

In our meetings, we plotted a full-fledged onslaught against the corporate advertising machine. We wanted to turn television into a battleground of competing narratives; fight an all-out guerrilla war where our mindbombs are constantly popping up all over the TV mindscape, neutralizing the corporate pro-consumption agenda. We thought that this kind of meme war would capture the imagination

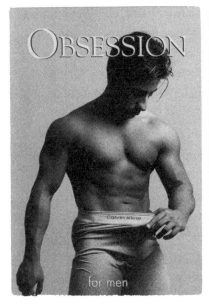 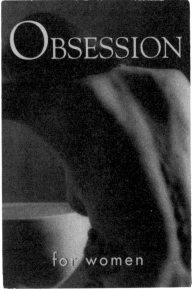

Obsession for men   Obsession for women

of the world, and we were confident that in the end, we would emerge triumphant.

We published a book called *Culture Jam*, which skirted the bestseller lists alongside Naomi Klein's *No Logo*. The movement took off. At its peak, the term became a meme in itself. "The studio for the cultural jammer is the world at large," the great experimental musician Don Joyce put it, and you saw that playing out big-time. People scribbled over bus-stop ads. They "liberated" billboards. They rubbed out the logos on their clothing and shoes and appliances. High-school teachers had their students make spoof ads and sent them to us.

Jesus, it was fun. We were playing jazz now, becoming a legitimate cultural counterforce. We'd cracked the code! We were doing… something. It was hard to tell exactly what. Or for whom.

But it was exhilarating suddenly to be standing toe to toe with corporate execs and sometimes landing a punch.

Like satire, culture jamming only *seemed* light-hearted. Underneath, for many of us, it was serious business about who creates culture. Will culture be spoon-fed to us top-down by corporations, or will We the People generate it from the bottom up?

By now advertising had mushroomed into a trillion-dollar-a-year industry, pumping hundreds of pro-consumption messages into our brains every day, colonizing every bit of literal and figurative space. Product placement in movies sparked outrage at first, but then became ho-hum invisible. Ads popped up in video games, golf holes and referees' armpits. Companies paid kids to carve their logo into their haircuts. One company wanted to put brand logos in space next to the moon. Advertising was transforming authentic culture into commercial culture and citizens into consumers.

We circled back round to Nike. We had a special beef with these guys. Paying millions of dollars to sports celebrities to dispense bogus cool to teenagers: that is graceless behavior.

30-second "Obsession Fetish" TV uncommercial.

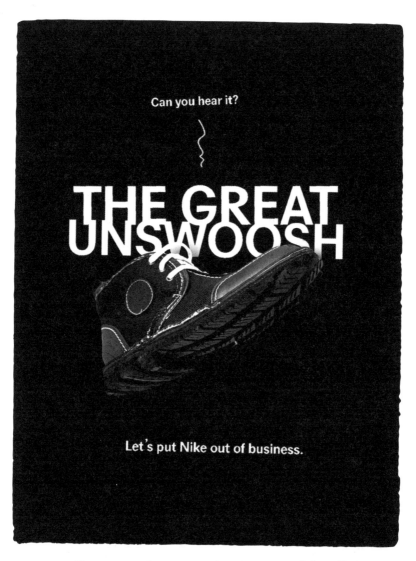

If we can uncool one corporation, we can uncool them all

    Nike would be a test case for a new move. We wanted to take on one of the giants not just in the marketplace of ideas but in the marketplace of… stuff. So we started our own shoe operation. The Blackspot Unswoosher: the take-no-prisoners pirate shoe, which we had ethically made in Portugal. *We think our brand can beat your brand.* Of course, Nike sold way more Air Jordans than we sold Blackspots. But then again, we weren't really interested in market share.

If our Blackspots could take even a fraction of a percent of business from Nike, then we'd have set a precedent that would inspire business-minded social activists to get involved in other areas — to become a virus at the heart of corporate capitalism.

We declared the Blackspot to be an open-source brand. Uncopyrighted and uncopyrightable. We wanted a nimble new breed of entrepreneurs to rise up — we called them "antipreneurs" — to launch Blackspot Coffee shops (with books & music, part library, part meeting place, part altar to caffeine). We wanted to revive the proud underground tradition of coffee shops as incubators of dissent. We'd sell fair-trade, organic coffee — revolution in every cup — and gradually push Starbucks out of our neighborhoods in the bargain.

TV Turnoff poster

We dreamed of a chain of Blackspot restaurants, selling only locally sourced food (a slow surprise in every bite), so tasty and easy they'd chase McDonald's out of the dinner plans of even the most harried parents.

We saw a world with Blackspot Cola, Blackspot Water, Blackspot Apparel, Blackspot Books, Blackspot Music labels. You'd even be able to get a Blackspot loan. It was anticapitalism in capitalist overdrive — all the profits plowed back into community projects. If you gave capitalism a conscience, this is what it would look like: small trumps big, local beats global, and shared, open-source brands roll over jealously guarded copyrights. Bit by bit, industry by industry, our little Blackspot would reverse capitalism's top-down dynamic and change the way people lived their lives.

Looking back, that level of idealism seems almost crazily naive. But that's also what was great about it. We didn't second-guess ourselves. Didn't get swamped by existential angst and the impossibility of pulling much of it off. We just reveled in the fuck-it-all spirit of it all.

These days, when network TV is pretty much a trifle in most people's lives, it's easy to forget how much power the networks once had over us. Television was the circulation system of the culture in the same way the Internet is now. For us, it was the holy grail. We'd get the leverage there to break the hold that advertising-driven culture had on us. We'd slip our mindbombs in there and fight a meme war. Put cultural power back into the hands of the people.

But something happened on the way to the revolution.

The TV stations refused to sell us the airtime. I had many angry phone calls with them, but there was no way they would take our few thousand dollars and risk angering their million-dollar sponsors.

15-second "Tubehead" TV uncommercial: The Product Is You!

The average North American consumes five times more than a Mexican...

... ten times more than a Chinese person and thirty times more than a person in India.

(the pig lets out a loud burp)

We are the most voracious consumers in the world... a world that could die because of the way we North Americans live.

Give it a rest. November 26 is Buy Nothing Day.

30-second "Burping pig" TV spot

The ABC network's commercial clearance boss Art Moore was defiant. "There's no law that says we have to air anything," he shouted at me on the phone. "We'll decide what we want to air or not."

NBC's Richard Gitter said, "We don't want to take any advertising that's inimical to our legitimate business interests."

And Robert Lowery at CBS finally put it straight about our *Buy Nothing Day* spot. "This commercial . . . is in opposition to the current economic policy in the United States."

An eerie sense of déjà vu set in. I was born in Estonia where during Soviet rule people were not allowed to speak against the government. There simply were no media channels for debating controversial public issues because the government did not want such discussion to take place. And here I was 50 years later in "the land of the free" and there was a lack of media space in which to challenge corporate agendas — you were not allowed to speak out against the sponsors.

We spent over $100,000 fighting a legal action against the Canadian Broadcasting Commission (CBC), arguing that every citizen has the right under the Canadian Charter to walk into their local TV station and buy 30 seconds of airtime for a message they believed in. The CBC fought us tooth-and-nail asking for our case to be dismissed. After 10 years of legal wrangling, the British Columbia Court of Appeal finally gave us a tiny opening. They

15-second TV mindbomb

ruled that Adbusters Media Foundation did indeed have a legitimate freedom-of-speech case under the Canadian Charter of Rights, and our legal action should be allowed to proceed. But by that time, we could no longer afford to pay the lawyers their $650-per-hour fees. We were exhausted and broke. Our dream of duking it out with megacorporations on commercial television was fading.

But it turns out that nothing is more tantalizing than something you aren't allowed to get your hands on. People wanted to see the banned ads, to know why they were rejected. Our culture jamming mindbombs began to circulate like wildfire.

And we made a lot of people question what television, the most powerful social communications medium of that time, was really all about if citizens can't buy airtime under the same rules and conditions as corporations do. After all, doesn't Article 19 of the Universal Declaration of Human Rights clearly state that: "Everyone

has the right . . . to freedom of opinion and expression; this right includes . . . the right to seek, receive and impart information and ideas through any media and regardless of frontiers"?

When you google "culture jamming," it's sometimes represented as a fad. Something that flared brightly for a decade in *fin-de-siècle* America. It was everywhere and then it was nowhere. Why? Was it because the TV networks refused to run our ads? Or because we lost our First Amendment legal battle? Or because the corporations themselves pivoted, dropping those monolithic, years-long campaigns that offered up such fat targets, in favor of quick-and-dirty hits tailored to the Internet age? Or maybe it was because culture jamming, for all its cheeky verve, wasn't actually solving anything. That's what some folks said. We set ourselves up as a

resistance community seeking to discredit and devalue consumerist culture and corporate rule, but at its core our jamming was better at tearing things down than building things up. We were vehemently anti-something, but not really pro-anything. We weren't able to articulate a set of values, meanings and alternate ways of seeing and talking about the world — which is what any successful oppositional culture must learn to do.

I don't entirely buy that.

Here's what I think happened to culture jamming: it never actually went away. It just evolved from a ruckus on the fringes of mainstream culture to part of mainstream culture itself. We are all culture jammers now. Everybody on social media is a cultural warrior. The Net is the jam. It scatters pulses of meaning and sends them on a collision course with other pulses.

Meme warfare is the next evolutionary phase of the people's resistance. This is where our metameme insurrection will take place. This is where the planetary endgame will be won or lost.

The True Cost Revolution

On the back cover of the first issue of Adbusters, we put a stern-looking Planet Earth pointing an accusing finger: *I Want You To Curb Your Consumption.*

The world was in a phase change. The Berlin Wall was about to fall. Solidarity was in the air, and the environmental movement was driving it. We'd seen the whole Earth from space and by God it looked vulnerable hanging out there in the darkness. Environmentalism wasn't some kind of consensus we all arrived at. It was a moment of divine reckoning. The Earth was alive. She was Gaia: the rivers were her veins; the forests her hair; the oceans her lungs. And we humans an organic part of her. It felt like we were embarking on a spiritual journey that would take all of humanity to a new level of consciousness.

Sure, there was a lot of scary stuff: vanishing forests, poisoned rivers, acid rain. Corporations behaving badly. Network television hijacking our attention with a barrage of pro-consumption messages every few minutes. But it was nothing we couldn't fix.

Then a curve ball out of the blue.

We *kind of* knew about global warming, but it was buried on our list of concerns, somewhere below pollution and the ozone hole. Now, suddenly, it was the whole show.

What we had on our hands, it became clear, was a terrestrial suffocation event. The oceans slowly starved of oxygen — an *I Can't Breathe* dirge for the planet itself. The Great Barrier Reef dying slowly and then quickly. Methane bubbling up from arctic permafrost. Species winking out at mind-blowing rates. The ecological structure that evolution had built up over billions of years was crashing down upon us.

This wasn't a cleanup job anymore. It was about survival.

Over the next few decades, the environmental movement mobilized — we came up with a flurry of ways to curb our emissions: We doubled down on carbon sequestration. Started massive kelp farms, harnessed streams in run-of-river hydroelectric projects, captured tidal energy at headland pinch points. We built wind farms and solar arrays all over the planet. We transitioned to electric cars. And many of us tightened our belts and learned to live more lightly on the planet.

But somehow, whatever we did never seemed to be enough.

Fifty years of environmental action has added up to little more than a rounding error on the climate emergency. "Cap and Trade" has been a bust. Carbon credits are the papal indulgences of the 21st century. Climate conferences are little more than kabuki theatre where world leaders sing their lines with great passion, then go home and continue doing business as usual.

It's totally baffling why we humans have been unable to come up with a plan for reducing our carbon emissions. Even more baffling is why we don't feel any urgency about that failure.

The late Stanford philosopher and therapist Paul Watzlawick had a good way of explaining how to get out of impossible jams. When change is required, there are different ways to think about the level of creativity that's needed.

A "first-order" change is to stamp on the gas pedal. A second-order change is to shift gears. A third-order change is to get out of the car and find another way to get there.

That's where we are now.

In our brainstorming sessions, one tantalizing idea kept popping up: If you dig deep into the innards of the capitalist algorithm, you discover a glaring flaw. It's that the vast majority of humankind's carbon emissions are *unpriced*. We buy a car for $35,000, then drive it around for ten years creating a thousand dollars' worth of global warming. Who pays for that damage? Do future generations have to clean up our mess? The illogic extends from the gas we pump into our cars to the smart phones we carry in our pockets to the Big Macs we wolf down at McDonald's. Out of the billions and billions of transactions made every day in our global marketplace, only a tiny fraction reflect their true cost. And each one drives us a little closer to global system collapse. With every bogus transaction, another drop of meltwater slides off an iceberg, another puff of CO_2 rises to the sky, another bubble of methane wafts up from the tundra. If we keep repeating that mistake, billions of times a day, week after week, month after month, year after year, what do you think will happen?

To date, only a handful of economists have bothered to think about the true cost of what we buy and do. They speak the language of efficiency and have taught the whole world to do the same. So why are so many of our leading economists silent, then, on these, the greatest inefficiencies of all? Why are our markets not telling the truth? Why are we selling off our natural capital and calling it income? Why is the profession of economics committing such a monumental system error?

Let's figure this out. What is the real cost of shipping a container load of toys from Chongqing to Los Angeles? Or a case of apples grown in New Zealand to markets in North America? And what is the true cost of that fridge humming 24/7 in your kitchen . . . that steak sizzling on your grill . . . that car sitting in your garage? What are the by-products of our way of living actually costing us? Grab a calculator and let's get at this. Instead of watching economists pontificate endlessly about interest rates, stock-market swings and GDP growth, let's put them to productive big-picture use crunching the real cost of things.

We start with the little stuff: plastic bags, coffee cups, paper napkins. Economists sleuth out the eco-costs — say it's five cents per plastic bag, ten cents per cup and a fraction of a cent per paper napkin — and those we tack on. We're already doing that with the various eco-fees and eco-taxes included in the price of tires, cans of paint and other products. But now, spurred on by the ever worsening climate crisis, we abandon the concept of ancillary fees and taxes and start implementing true-cost pricing right across the board.

TRUE COST PLASTIC

After raiding nature's warehouse of wood and stone and metal, we turned, in the early 20th century, to plastics. Here was something cheap and strong to build a space-age world.

Now that miracle invention is choking our landfills, polluting our rivers and oceans and poisoning our bodies and food chains. The Organization for Economic Cooperation and Development predicts plastic use will nearly triple by 2060. Canada and the European Union have banned single-use items and some activists and scientists are advocating capping and reducing plastic production and use. But it's obvious that none of this will be nearly enough to fix the problem.

Here's a strategy that will.

Economists do the research and come up with their best estimate of the environmental and health price we pay — say it's $500 per ton. Every manufacturer, corporation and retailer that uses plastic in their business will then be required to account for that. Maybe it's a surcharge of 25 cents on every bottle of Coke. If Coca Cola can't take a hit like that on their margin, they'll have to change their business model. Likewise, the automobile industry will have to redesign their

cars. Food producers and supermarkets will have to adapt. Every business that uses plastic will have to adjust their business model.

The cost of living will go up, and that'll hurt. But plastic packaging will gradually disappear from our lives. We'll buy our groceries in paper, cardboard and glass containers. We'll wash our plates, knives and forks and use them year after year, some for a lifetime. The garbage gyres in the oceans will shrink and finally disappear. Microplastics will stop plugging the tissues and brains of every mammal including us. And the nightmare of bringing up our children in a world awash in plastic will slowly fade away.

TRUE COST DRIVING

Once we add on the environmental cost of carbon emissions, the cost of building and maintaining roads, the medical costs of accidents, the noise and the aesthetic degradation of urban sprawl, your gasoline powered automobile will cost you around $100,000, and a tank of gas $350. You'll still be free to drive all you want, but instead of passing the costs on to future generations, you'll pay up front.

Plenty of people will dismiss the concept as unrealistic and dangerous. There will be howls of protest. Traditional lefties will point out how true cost would create a two-tier regime in which the ultrarich can afford to emit as much frivolous CO_2 as they like, while for the bulk of humanity everyday life will be more miserable than ever. Politicians will dismiss it as electoral suicide. Industries will lobby vehemently against it — at least in the beginning. But as the planet heats, and all the other strategies have failed, true cost may turn out to be the only way left to avoid total climate catastrophe. And once it kicks in, we'll see a fundamental transformation in how we get around on our planet. Car use will plunge. Ride sharing and bicycling will spike. People will live closer to work. Demand for monorails, bullet trains, subways and streetcars will surge. A paradigm shift in urban planning will calm the pace of city life. Cities will be built for people, not cars. Our skies will be clearer. Breathing easier. Minds calmer. The specter of the climate emergency will no longer preoccupy our every waking moment.

TRUE COST EATING

We estimate and add in the hidden costs of our industrial farming and food processing systems. That chicken that was never allowed to spread its wings and walk outside will cost you $50. The price of imported groceries will include the true cost of shipping them long distances. An avocado from Mexico will cost you $25. You won't be able to indulge so often. And that shrimp from Indonesia? Once the eco-devastation of mega farming and container shipping are added on, it'll run you two or three times what you're paying now. A Big Mac will cost a lot more. So will most meats, produce and processed foods. You can still eat whatever you want, but you'll have to pay the real price.

It will be tough at first, especially on lower-income families. But the cost of organic and locally produced food will fall and provide a good alternative. Local farmers will be celebrated. We'll grow tomatoes on our verandas, eat at home more and maybe lose some weight and be a little healthier. Bit by bit, purchase by purchase, lifestyle change by lifestyle change, our diets and food systems will creep toward sustainability.

TRUE COST SHIPPING

For years it's been ridiculously cheap to use mega tankers to ship container loads of stuff across oceans. Much of that will stop. Our current way of exporting and importing goods, the one economists have been touting as a way to spur growth, but which depends on a mightily subsidized ocean transportation system, will no longer fly. The cost of all imported items at Walmart, Amazon and multinational megamarts will soar. The whole tenor of world trade will heave. Exports and imports will stabilize at a reduced level. Billions of purchases every day will come back to your neighborhood. Globalization — capitalism's bred-in-the-bone burden — will cease to be the dominant economic paradigm. Modest fleets of cargo ships will go electric, or sail around the world on wind.

TRUE COST SEARCHING AND SCROLLING

What we call "the cloud" is actually 8,000 data centers peppered throughout the world, and they gobble an enormous amount of energy every year.

According to a recent study, a typical Google search from a desktop computer generates about 7g of CO_2. Two Google searches generate as much carbon as boiling a kettle for a cup of tea. Big Tech downplays how much energy their servers use, but make no mistake, the demands are going to go through the roof as everyone starts putting machine learning to personal use.

Since the beginning of the digital revolution, we've largely ignored the real cost of our online lives. But now the time has come to rethink our scrolling like we've already rethought our eating, driving and flying. In other words: stop doing it mindlessly. Stop doing so much of it, and let's figure out how to calculate and internalize the costs.

NEXT LEVEL ACCOUNTING: THE SOCIAL AND PSYCHIC COSTS

You're cruising along an eight-lane highway and suddenly everything lurches to a halt. There's a lot more going on here than a hefty blast of carbon wafting into the air. A traffic jam is a huge collective stress event. There are health costs to being pinned in your car, on a dammed river of steel, fingers tightening on the wheel, blood pressure rising. Mental health costs too. A recent Swedish study found that a daily commute of forty-five minutes increases your chance of divorce by 40 percent.

The psychic costs that our current system imposes on us are horrendous, and we're just at the very early stage of realizing how devastating they really are.

What is the psychic cost of advertising, that daily broadside of pro-consumption messages that's pickling your neurons? Or the mental toll of obsessively checking your phone — basically tugging on the leg-hold trap of Big Tech's surveillance algorithms, over and over and over? Or the psychological damage of *urgency*, the punitive ticking clock that every link in the supply chain and every component of the gig economy runs on? Or the social solidarity cost of losing most of the indie shops in your neighborhood as Starbucks, Domino's and Home Depot muscle their way in. Ask yourself, what's happening to your soul when you're walking like a zombie in a mall at Christmas with *Silent Night* playing in the background? All this is part of the True Cost story — and so must eventually be part of the accounting — of the epidemic of mood disorders, anxiety, loneliness and depression now sweeping the planet.

For conventional economists, True Cost is a frightening, heretical concept. Once implemented, it will reduce the flow of world trade, curb consumption and slow growth. It will force economists to rethink just about every axiom they've taken for granted since the dawn of the industrial age.

The efficiency of size will be challenged. The hidden cost of Walmart coming to town, revealed. The logic of never-ending growth on a finite planet thrown back in economists' faces.

"Progress" itself will be redefined.

There'll be angels-on-a-pin debates about the psychological and social costs of our most sacred values like unfettered freedom and hyper-individuality. A fascinating new branch of economics called *psychonomics* will emerge.

Transitioning towards True Cost will be the most challenging and disruptive economic / social / cultural project we have ever undertaken.

But it will also be magically transformative.

In a True Cost world, there'll be no need for pleading and hectoring, no need to wallow in conflicting consumerist emotions. No one will be badgering you to eat less meat. No one will make you feel guilty about owning a car, or for going on that holiday to the Bahamas. None of that. All you're being asked to do is become a consumer in a new kind of marketplace.

Instead of "lowest price wins and don't ask too many questions," market forces will align in surprising new ways. You will become part of a worldwide process in which every one of the billions of market transactions made every day are working *for* rather than against us.

In the current language of economics, the costs you don't see, the ones that don't show up in the models, are dismissed as "externalities" — just the trims and ends left over when you run the growth numbers. Only a handful of economists have bothered to think of these costs as anything other than marginalia — a few paragraphs in Gregory Mankiw's *Principles of Economics* textbook.

True Cost will put a shine on "the dismal science." Economists will suddenly have a crucial purpose in life: to calculate and fold in all the costs of our way of doing business. This will ground them and give them something real to do. It will create a virtuous,

ULTIMATUM #1 TO WORLD LEADERS

Our planet keeps heating, hope for a livable future fading, and you are doing next to nothing to stop it. You are botching the biggest existential crisis humanity has ever faced.
So we, the people, are stripping you of your command. From now on you answer to us. Here's what we want you to do:

Declare a Global Climate Emergency.

Then:

Halt all subsidies to oil companies.

Eliminate all tax havens.

Impose a 1% Robin Hood Tax on all stock market transactions and currency trades.

Revoke the charters of corporations that commit heinous crimes (ExxonMobil, Wells Fargo, Purdue, et al). Simply wipe them off the face of the Earth.

Move relentlessly towards a global marketplace in which the price of every product tells the ecological truth.

This is not a conversation anymore. It's an ultimatum. If you don't move on our demands, then every Friday, in cities across the world, we will ramp up the intensity of our civil disobedience and bring this doomsday machine of yours to a sudden shuddering halt.

The Third Force

forward-looking occupation out of a retrograde one. The profession will be a highly desirable path to pursue — something a young grad is proud to devote their life to. Environmentally minded students will stream into Econ 101, because now it incorporates sociology, anthropology and psychology. It is the *Queen of the Sciences,* the essential discipline for nudging our global system onto a sustainable path.

HOW TO DO IT

The Universal Product Code (UPC) is the perfect mechanism for implementing True Cost. Almost every product sold around the world already has a UPC code on it. It was originally created to speed up the checkout process and track inventory at grocery stores. Now it becomes an essential part of the True Cost project. The human experience of buying something is transformed. When you swipe it, a True Cost price adjustment automatically kicks in. All the costs of making and marketing and shipping and distributing the product you're buying are baked into the price. Adam Smith's invisible hand

The True Cost Party of America

OUR PLATFORM

Immediate implementation of a 1% Robin Hood Tax on all stock-market transactions and currency trades

Radical curbs on derivatives and credit default swaps

A 24-hour rule on flash trading — you buy it you keep it for 24 hours

No corporation is allowed to hold more that 25% of the market in any industry

A three-strikes-and-you're-out law for repeat corporate offenders

A move towards a True-Cost global marketplace in which the price of every product tells the ecological truth is suddenly being used in the most profound way to adjust your purchasing behavior. Bit by bit, you accept the idea that you should always pay the True Cost of everything you buy.

One swipe, one truth. Sticker shock: take it or leave it.

"It'll never fly," say many of the environmental luminaries we've put this to. It's just way too radical and impractical.

Nothing of this scope, on this scale, has ever been tried. To most people, it feels like about Plan D — after all the more 'sensible,' green-energy, carbon tax and techno options have been kicked around.

Just on the level of human psychology, True Cost is a big, big ask. Self-indulgence is the kingly spoils of consumer capitalism. We *love* our dishwashers, spin dryers and airflow toasters. We love our throw-away diapers, pre-washed veggies and ready-to-eat dinners delivered right to our door. And we love cruising in our automobiles with our favorite music playing. The idea of giving some of that up, of living tougher, more austere lives is anathema. We're talking about curbing the personal freedoms we've taken for granted for centuries. "Dammit, my right to pig out, play hard and put pedal-to-the-metal is constitutionally protected!" Americans will say. Anything less is heresy. We can't do this. We won't do it. Go to hell!

But as the planet heats, the mood will change. Ecological collapse is a slow-motion catastrophe. You don't feel it yet, you cannot grasp the urgency of it. Because your hair isn't on fire. *Yet*. But once we pass a tipping-point — and we'll absolutely know it when it happens — when resource skirmishes erupt into full-scale battles, and slow violence turns into fast violence, and suddenly it's *your* children who are hungry and *your* house that's being swept away and *your* country that's failing . . . that's when you'll forget "It'll never fly" and reach for the ax on the wall.

Recently some climate activists have begun to openly contemplate the possibility of directly sabotaging the infrastructure of the carbon economy. Foremost among them is the academic

Andreas Malm, whose book *How To Blow Up a Pipeline*, calls for smashing the tools of fossil-fuel extraction as a last-ditch means of averting ecological collapse. "Damage and destroy new CO_2-emitting devices, put them out of commission, pick them apart, demolish them, burn them, blow them up. Let the capitalists who keep investing in the fire know that their properties will be trashed."

I have no problem with destroying machines and property as long as people are not harmed. And that may well be the direction environmental activism will go. But it's not enough. We also need a long-term strategy, a *process*, like True Cost, that gradually bends our global system towards sustainability.

Here's how it could unfold:

We seed the global imagination with our memes and incitements. We get in the faces of the current flock of neoliberal economists running the world economy. *True Cost Pricing* is our creed. We push it relentlessly.

We explode 15-second mindbombs on business news channels and the evening news.

We insert True Cost into the platforms of Green Parties around the world and work to unite them into a unified global force.

We give birth to the *True Cost Party of America*.

As global temperatures soar and summer heatwaves become unbearable, we unleash waves of civil disobedience that bend world leaders to our will.

Our planet is now a Pachinko ball tumbling through the universe. There is simply no predicting the outcome. It could well be that the best efforts of our scientists, political leaders and activists will come to naught, and humanity will spiral into a new dark age that beggars the imagination. But it's equally possible that when all the other ideas have failed and the world suddenly tips into panic mode — when all hope is lost — then an 11th hour desperation will open a crack in the human psyche. And that will be the moment we unite behind a monumental communal project: a radical turn toward a True Cost future.

Jan. 19		2029

An almighty brainstorm is raging
... everybody is asking themselves
the same question: What the fuck
happened ... why did we crash ...
why did the center not hold?
Was it because advertising pickled our
neurons? ... or because the prices of all the
stuff we bought never told the
eco-truth? ... or because flash bot trading took
over and distorted everything? ... or maybe
because we allowed secrecy to creep into every
nook and cranny of our political life? ... or
maybe the dumbest, stupidest mistake of them
all: granting legal personhood to corporations?

An insatiable desire now to tinker with all these toxic remnants of the old world order.

Pin up in the corridors

Battle for the soul of economics

In 1987, the National Film Board of Canada gave me a bunch of money to make a one-hour documentary about the global economy. This was pretty sweet. I'd get to travel the world, talking to business leaders and economists and economics profs — some of the most visionary minds shaping fiscal policy, really the managers of our daily lives.

I started with the rock stars. Men like Walter Williams, from George Mason University, who boasted that economics had become a kind of master discipline. "We've established some very, very solid laws that are similar to those in physics..."

Economics was unfolding, they insisted, just as Adam Smith said it would when he created it two centuries earlier. JM Keynes — a mathematician who designed the big macroeconomic levers that just about every Western economy would rely on — likened economics to dentistry: a nuts-and-bolts profession that simply got the job done: no guesswork, no pain.

I remembered then what had attracted me to this subject in the first place. That kind of swagger often comes from a place of weakness, not strength. When something's being aggressively oversold, that often means it's broken. I had a nagging feeling about economics: there's no *there* there.

Neoclassical economists basically say: Trust the models. Trust us. We have this thing figured out. We can micromanage growth, engineer prosperity, create jobs and keep things humming. For them economics is just an exercise in pure logic.

Then I ran into Hazel Henderson. She wasn't an economist, but a writer. Her husband was an economist. So she wasn't in the economists' clubhouse but close enough to hear everything through the door. And what she heard struck her as mostly bogus. "Economics isn't a science," she declared. "It's just politics in disguise."

Henderson, who would go on to co-invent an ethical investment strategy called Biomimicry Finance, gently nudged me in an exciting new direction.

I started to investigate alternatives. And quickly stumbled on a story of resistance and growing discontent a hundred years in the making.

Nicholas Georgescu-Roegen. Frederick Soddy. Kenneth Boulding. Robert Heilbroner. E.F. Schumacher. Herman Daly. These were seriously out-of-the-box thinkers. They had creative ideas about how the world really works. They were dismissed as gadflies by people who didn't understand what they were talking about: *entropy, thermodynamics, chaos theory*. Each in their own way, and with a depth and profundity I found exhilarating, questioned what we all thought were the sacrosanct axioms of economic science.

Are the costs of our way of doing business being properly tallied?
Do we know how to measure progress?
Can we keep growing?

These folks thought wide. They thought wobbly. They thought *through* the WASP-y arrogance of neoclassical economics' straight-line ambitions.

It's not like they were anti-science. Soddy won the Nobel Prize for chemistry. So he was in a good position to notice that what was going on in the economics department didn't quite add up. That there was no real science going on in the sandbox the economists had built for themselves.

"Stop creating money out of nothing," he told them.

"Growth may cost more than it's worth," Herman Daly warned them.

"Existence is a free gift from the sun," Georgescu-Roegan reminded them. (When they scratched their heads about that, he doubled down with: "Waste increases in proportion to the intensity of economic activity.")

These mavericks floated tantalizing new ways to think about markets, growth, value and progress. They were playing a longer game, a more cosmic game. A more beautiful game.

I returned home to Vancouver convinced that my initial hunch had been right: there is something seriously wrong at the heart of economics. But I also felt now, despite my earlier misgivings, that economics really does matter.

That economics is destiny.

Get it wrong and people revolt, ecosystems collapse, empires fall.

But get it right — even for a brief moment — and a bolt of optimism, creativity and enlightenment surges through the land.

If economists could see past their mathematical models and formalist pretensions and embrace psychology, sociology and anthropology — even history and religion — their discipline could evolve into a catch-all science that could solve many of the ills that plague humanity.

My documentary aired a couple of days after the stock market crash of 1987. None of the mainstreamers had predicted this event. None had the foggiest idea why it happened. I expected my film to trigger a national debate about the flawed axioms at the heart of the profession. But the moment passed and ... nothing happened. It was puzzling and maddening all at once. I felt like the kid sitting in the back of the class with his hand up but the teacher never calls (because the kid is, I guess, annoying).

For the next few years at *Adbusters* we kept the drumbeat. We interviewed Herman Daly, ran excerpts from his anthology *Towards a Steady State Economy* and eventually named him our *Man of the Year*.

We also produced a cheeky 30-second TV spot about economists' disfunctional accounting. Atop the images of clearcut forests, oil spills and cancer patients, a voice-over says:

For years economists have defined the economic health of a country by its Gross Domestic Product.
Trouble is, every time a forest falls, *the GDP goes up.*
With every oil spill, *the GDP goes up.*
Every time a cancer patient is diagnosed, *the GDP goes up.*
Is this how we measure economic progress?
And then the outshot:
ECONOMISTS MUST LEARN TO SUBTRACT!

Our plan was to keep running it on business news channels until it hit a nerve. We relished the prospect of a public debate about how economists measure growth, define progress and generally run things. We were confident we could win the debate. But our TV mindbomb made no impact at all — because it was never seen. None of the TV stations in the US and Canada would sell us airtime.

THE STUDENT UPRISING
There's a word for people who focus obsessively only on what matters to them, in such granular detail that they lose sight of the big picture and forget that what they do affects other people and other things, and that not everything needs to happen *right now.*
That word is *autistic*.
It's the word 15 economics students used at Paris's École Normale Supérieure when they stormed out of their classes in the summer of 2000 and organized protests and teach-ins. They were talking about a cultural disorder, not a neurological one.
The reigning neoclassical model of economics is *autistic*, they said. It's so rational it's *ir*rational. They demanded widespread reforms within economics teaching, which they said had become

enthralled with complex mathematical models that work only in conditions that don't exist.

The students were rebelling against the straw man at the center of economic science — a pathetic parody of a human being called the "Rational Utility Maximizer." This guy runs around making perfectly predictable choices within perfectly functioning markets. He's never depressed, never sickened by pollution, never emotional, never a dreamy wanderer, he never falls in love.

But of course, real humans are not like that. We fly off on wild tangents, have selfish and altruistic impulses, do crazy things. We feel guilty when we overindulge, get depressed when we lose our jobs, seek revenge when people do us wrong, and we often refuse to buy something just because our values don't jive with those of a corporation. And we routinely thwart our "rational self-interest" by pulling off remarkable feats of compassion and teamwork.

Neoclassical economics cannot handle this truth. It has achieved its coherence as a "science" by amputating most of human nature.

Three Nobel Laureates — Wassily Leontief, Ronald Close and Milton Friedman — expressed similar concerns. "Existing economics is a theoretical system which floats in the air and bears little relation to what happens in the real world," complained Close. And near the end of his life, Milton Friedman conceded that "Economics has become increasingly an arcane branch of mathematics rather than dealing with real economic problems." The only thing that matters in science is predictive power, he said. Your models need to get better and better, in a tightening circle of accuracy. If they don't, it's not science. Economists roundly failed that test.

None of these warnings had any impact whatsoever on economics curricula. Mark Blaug wrote: "We have created a monster that is very difficult to stop."

But the students tried. They called their movement autisme-économie — "post-autistic economics." These were some of the brightest students at one of the top schools in Europe. They'd done their homework, so they were equipped to handle the blowback from the old guard (like Robert Solow from MIT, who called them "children").

Networking over the still-budding internet, the students spread the message that economics has narrowed to the point where it cannot contend with real-world challenges like inequality, financial instability and climate change.

Their rebellion surged after an open letter signed by a thousand of them was published in *Le Monde*. In class and in the corridors, they waved signs: "We wish to escape from imaginary worlds." They mocked their professors' "uncontrolled use of mathematics," number-crunching as an end in itself. Where in our curriculum, they asked, are the alternative perspectives — the hundreds of credible intellectuals and freethinkers who offer fresh ways to explain the world, through the lens of feminism, socialism, behaviorism, even Buddhism? Where is the syncretism that's the sign of an actual grown-up mind?

"Wake up before it's too late!" the students shouted.

Their rebel spirit spread to Cambridge and Harvard and dozens of universities around the world. A slogan splashed on a wall on a Madrid campus summed up the disconnect: *"¡La economia es de gente, no de curvas!"* — "Economics is about people, not curves!"

The post-autistic movement had its shining moment . . . and then, like other movements before it, it faded away. But it left an indelible mark in the minds of a generation of students. It was a seed that could grow.

> Before economics can progress, it must abandon its suicidal formalism.
>
> — Robert Heilbroner

Bad Karma

In high school the cool kids smoked. So I started smoking too. And I continued to smoke in university. I just changed brands. Gaulois were too pretentious; I went for British brands like Peter Stuyvesant, with its pure white pack.

Then I started to hear murmurings: cigarettes cause lung cancer.

It was still just a rumor — at least that's how the industry spun it. "The link hasn't been proven," said Philip Morris. PM had marshaled a team of corporate lawyers and PR flacks — not to mention publicly skeptical doctors on their payroll. The evidence just isn't there, they said. The average smoker's chances of getting lung cancer from cigarettes is roughly the chance of being struck by lightning.

So I kept puffing.

But I tried to quit, again and again. Problem was, I couldn't edit a film without smoking. You can imagine it: you're immersed in the flow, the rational part of your brain is in park, the wild reactive part firing on instinct, and your hand instinctively reaches for a dart. For me back then smoking was woven into the ritual of doing creative work on a deadline.

The magic of tobacco is that it's both a stimulant and a relaxant. The smoke cloud itself is hypnotic.

One morning, after a lot of boozing and chain-smoking through many nights of editing, I woke up feeling like shit. And I stopped smoking. Again.

But this time I stayed clean. One, two, three days. On day four I woke up feeling . . . amazing. Clear skies.

Not long after, I ran into one of my mother's neighbors. He looked awful. I learned he was dying of lung cancer. He had only a few months to live. I looked in his eyes and saw that he had given up. I'd gotten lucky; many hadn't.

It dawned on me that Philip Morris did this. They knew they were killing people and they covered it up. They built their empire on a business plan to sell toxic addictive products to people who didn't know any better. They were literally getting away with murder and had been for decades.

That's the moment I started to hate corporations.

"Hate" may be the wrong word. What do you say about someone who betrays your trust? It sounds insane now, but back then people actually *did* trust corporations. You took it for granted that they were acting in the public interest. They were synonymous with progress, and faith in them seemed as sensible as faith in anything else.

The more that emerged about Philip Morris's duplicity, the more I could feel the red mist of rage clouding my vision. Those lying bastards. Just so diabolically calculating. Hook the customer early and squeeze a lifetime's worth of profits out of them — and when regulators close in, you shift your focus to the developing world where the rules are looser and you target those kids.

I thought: How can a company like that even be allowed to exist?

That rage bubbled on the back burner for years. I didn't really know what to do with it. Until the day in 1991 when Richard Grossman walked through the door of *Adbusters*.

He asked if I had a few minutes to chat.

Richard was a lawyer and a legal researcher. He'd spent the last decade immersed in American history, connecting the dots of corporate ascendance. He knew where the bodies were buried.

Richard spoke softly, with the quiet authority of someone who has done the work. There was something about him that was totally free. He was tracking to true north, his mission as tightly defined as marching orders, and as large as the future of democracy itself.

Corporations, he said, were "worms in the body politic." (He talked like that.) They are nothing more than "legal fictions" and We the People need to redefine our relationship with them. We have to change the way we speak and even think about them.

People have no idea how insidious corporate culture is, Grossman said. It sculpts the personalities of consumers, encouraging the very traits Americans claim to hate: self-censorship, low aspirations, soul-crushing conformity.

And then he told me a tale that would change my life and work at *Adbusters*. Call it The Unofficial History of America.

The official history of America is one every kid knows. It's a story of fierce individualism and heroic personal sacrifice in the service of a dream. A story of hungry settlers carving a home out of the wilderness. A story of a revolution, beating back British imperialism and launching a new colony into the industrial age on its own terms.

It's a story of America triumphant. Of its rise after World War II to become the richest and most powerful country in the history of the world, the land of the free because of the brave, an inspiring model for the whole world to emulate.

The unofficial story of America is quite different.

It begins the same way — in the revolutionary cauldron of colonial America — but then it takes a turn. A bit player in the official story becomes critically important. This player turns out to be not only the provocateur of the revolution, but in a sense its saboteur.

The corporation.

We tend to think of corporations as a fairly recent phenomenon, the legacy of the 19th-century robber barons. But they were front and center in pre-revolutionary America. They had a beachhead before the pilgrims landed at Plymouth Rock; the language of their charters is in the DNA of the Constitution. The Virginia Company of London, the Massachusetts Bay Company, the British East India Company: these were the Amazon and Wal-Mart and Google of their day.

But back then our dealings with these behemoths were very different. The colonials were highly suspicious of them. They recognized the way British kings and their cronies used them as robotic arms to control the affairs of the colonies, to plunder resources and bring them to the motherland.

The colonials pushed back. When the British East India company slapped duties on its incoming tea (and they had the colonials over a barrel; that's how monopolies work), well, you know that story. Radical patriots protested, ships were turned back at port and 342 chests of tea ended up in the saltchuck.

The Boston Tea Party gave young America a glimpse of its own strength. Self-determination was starting to seem not just possible but inevitable.

The Declaration of Independence freed America not just from British rule but also from the tyranny of British corporations, and for a hundred years Americans remained deeply wary of corporate power. Corporate charters were granted very selectively and for a specific purpose. Limits were set on how big and powerful a corporation could become.

Even the enormous industry trusts proved no match for the state in those early years. Antitrust legislation kept their appetites in check. Corporations couldn't participate in the political process. They couldn't buy stocks in other corporations. If they stepped out of line, the punishment could be severe. Charters could be revoked and some were. In 1832, the state of Pennsylvania found ten banks to be acting against the public interest and wiped them out of existence.

The people — not the corporations — were in control.

So what happened? How did corporations come to wield more power than the individuals who created them?

The shift began in the last third of the nineteenth century — the start of a great period of struggle between corporations and civil society. The turning point was the Civil War. Corporations made huge profits from procurement contracts. They took advantage of the disorder and corruption of the times to buy legislatures, judges and even presidents. President Lincoln foresaw terrible trouble. "Corporations have been enthroned," he warned, shortly before his death. "An era of corruption in high places will follow, and the money power will endeavor to prolong its reign by working on the prejudices of the people . . . until wealth is aggregated in a few hands . . . and the republic is destroyed."

Nobody listened. Corporations continued to roll. They had the laws governing their creation amended. State charters could no longer be revoked. Corporate profits could no longer be limited. In hundreds of cases judges granted corporations minor legal victories, conceding new rights and privileges.

Then came a legal event that would not be understood for decades (and remains baffling even today), an event that would change the course of American history. In *Santa Clara County vs Southern Pacific Railway*, a dispute over a railway route, the U.S. Supreme Court deemed that a private corporation was a "natural person" under the U.S. Constitution and therefore entitled to protection under the Bill of Rights. Suddenly, corporations enjoyed all the rights and sovereignty previously enjoyed only by people.

Santa Clara was nothing less than a corporate takeover.

It's still widely considered one of the great legal blunders of the nineteenth century.

Corporations employed 80 percent of the workforce and produced most of America's wealth. They had become too powerful to legally challenge.

Many of the original ideals of the American Revolution had been quashed. America was being ruled by a coalition of government and business interests. The shift amounted to a slow-motion coup d'état — a gradual subversion and takeover of the institutions of state power. Except for a temporary setback during Franklin Roosevelt's New Deal (the 1930s), the U.S. has since been governed as a corporate state.

Today, corporations have nearly all the same rights as people: freedom of speech, freedom of the press, religious liberty, due process, freedom from unreasonable searches and seizures, the right to counsel and the right to trial by jury. They freely buy each other's stocks and shares. They lobby legislators and bankroll elections. They manage the flow of information, set our economic and cultural agendas and grow as big as they damn well please.

We the People have lost control. Corporations, these legal fictions we ourselves created two centuries ago, now rule over us. And we accept this as the normal state of affairs. We go to corporations on our knees. Please do the right thing we plead. Please don't cut down any more forests. Please don't pollute any more lakes and rivers (but please don't move your factories and jobs offshore either). Please don't use pornographic images to sell fashion to my kids. And please don't addict them to your algorithms. We've spent so much time bowed down in deference we've forgotten how to stand up straight.

The unofficial history of America, which continues to be written, is not a story of rugged individualism and heroic personal sacrifice in the pursuit of a dream. It's a story of democracy derailed, of a revolutionary spirit suppressed, and of a once-proud people reduced to servitude.

Richard Grossman had literally been pacing around the office as he laid all this out. "Two narratives," he said. "The difference between them is the difference between a country that pretends to be free and one that truly is."

"It's time to make a tactical leap, change the way we grant charters to corporations and take our country back!"

On that day, this mild-mannered visionary set a new agenda for our fledgling NGO and beyond.

Corporate Crackdown!

A corporation has no heart, no soul, no morals. It cannot feel pain. You cannot argue with it. That's because a corporation is not a living thing but a process — an efficient way of generating revenue. It takes energy from the outside (capital, labor, raw materials) and transforms it in various ways. In order to continue "living" it need meet only one condition: its income must equal its expenditures over the long term. As long as that happens it can exist indefinitely.

When a corporation hurts people or damages the environment, it will feel no sorrow or remorse because it is intrinsically unable to do so. (It may sometimes apologize, but that's not remorse — that's public relations.) Buddhist scholar David Loy put it this way: "A corporation cannot laugh or cry; it cannot enjoy the world or suffer with it. Most of all a corporation cannot love." That's because corporations are legal fictions. Their "bodies" are just judicial constructs, and that is why they are so dangerous.

We demonize corporations for their unwavering pursuit of growth, power and wealth. Yet let's face it: they're simply carrying out genetic orders. It's exactly what they were designed — by us — to do! That's why trying to rehabilitate a corporation, to urge it to behave more responsibly, is pointless. The only way to change the behavior of a corporation is to recode it, rewrite its charter, change its DNA.

When the verdict in *Citizens United vs the Federal Election Commission* came down in 2010, I remember thinking: *We're fucked.* CEOs know they are off the hook. No matter what they do now they are untouchable. And sure enough, an almost pathological arrogance set in in corporate boardrooms across America.

A corporate crime wave was unleashed. General Motors kept secret a faulty ignition switch that was killing its customers. Wells Fargo raked up record profits by defrauding millions of its customers. The Sackler family carried out a Machiavellian marketing scheme that killed millions. The mindlords of Big Tech transformed a whole generation of children into doom scrolling addicts. And Exxon Mobil continues to perpetrate its crime against humanity with impunity. In the stories of corporate malfeasance that I read about in *The New York Times* almost daily, there's a lot of emphasis on the multi-million dollar fines that corporate criminals have to pay, but the possibility of revoking their charters and wiping them off the face of the earth is never seen as an option.

In Kazuo Ishiguru's novel *Never Let Me Go,* cloned humans are raised as organ donors. The clones are gentle and sheeplike and resigned to being slaughtered before their thirtieth birthday, their parts harvested to keep the elites alive. They never band together and try to rise up because it never occurs to them that they deserve more. There's no solidarity in this group. They simply bear their lot as sorry, defeated individuals. Many of us are like those clones. We think of ourselves as precious individual snowflakes, rather than cordwood for the corporate furnaces. That is our mistake.

Somewhere between *Santa Clara* and *Citizens United*, we lost the plot. We lost our dignity. We allowed corporations to dominate us. How we let this happen no longer really matters. The question is: What are we going to do about it? Can we come up with some bold strategic moves potent enough to upend the top-heavy capitalist paradigm we're currently trapped in?

THE 25% RULE

Today Google controls 90% of Internet searches. Amazon commands more than half of online purchases. Five corporations handle a quarter of the world's oil. Seven grain traders supply half the world's food. Most industries are more clotted today than in the time of the robber barons. Airlines, telecom, banking, pharma, hospitals, agribusiness, waste management: a handful of big players dominate them all.

How can we live dignified, independent lives knowing that the food we eat, the coffee we drink, the medicines we take, the news we consume and the internet we spend so much time on, are each controlled by a handful of mega-corporations?

We're caught in a maddening, demeaning, insufferable situation, unworthy of freedom-loving people.

Bust them up!

Except you can't. If you threaten their supremacy in any way, swarms of lawyers, lobbyists and million dollar PR campaigns are immediately deployed to restore order. They spin irresistible "consumer welfare" stories that are almost impossible for civil society to counter.

And mergers are hardly ever turned down. Antitrust laws are useless. We've tried this experiment for decades and no one thinks it's working.

Let's face it: governments will never dissolve mega-corporations. It's up to us!

We, the citizens of the world must make a blanket commitment: *We don't want corporations to dominate us.* In every corner of our lives, we will support the "creative destruction" of every mega-company that has too much power.

One tantalizingly simple way to wrest back some of our sovereignty is to pass a law that prohibits any corporation from having more than 25% of the market share in any industry.

One quarter. That's all you get!

This will immediately be dismissed as outrageous. Are you crazy? That's un-American. It's a slap in the face of everything our free-enterprise system stands for!

But think about it. Would we be in an existential climate crisis today, if we had cut Shell, BP and ExxonMobil down to size fifty years ago? Would the 2008 financial meltdown have devastated the world economy if Goldman Sachs, JPMorgan Chase and Morgan Stanley were a fraction of their size?

And where would the internet be today if we had never allowed Google, Facebook and Amazon to run amok in their domains? Would surveillance capitalism even be a thing? Might the original dream of a crafty, decentralized FREENET have survived instead? Would facts, truth and democracy still mean something? Would our sons and daughters be a little less anxious and depressed? Could Brexit have been prevented; populist movements forestalled? Would Trump have been kicked to the curb? Would the world be as polarized into alternate realities as it is today?

As surveillance algorithms tighten their grip on us and global temperatures soar, there will come a moment when We the People cry "Enough!" We'll have a collective aha moment and realize that we will never win the planetary endgame if we allow megacorporations to keep calling the shots and lording it over us. That's the moment we'll find the courage to think the unthinkable, demand the impossible and cut corporations down to size.

And when it happens, a tsunami of entrepreneurial zeal will immediately be released. What Schumpeter called "creative destruction" will explode in every industry, in every corner of our lives, clearing the way for a more diverse, more dynamic and sustainable business culture.

A SIZE TAX

I'm old enough to remember when the first supermarket came to my neighborhood in Adelaide, Australia, and ran the local butcher out of business. Weeks later I saw him walking into that store. He was working there now, in the meat department. On one level he was probably glad to have a job. But the new terms were written on his face. He looked so sad. The before-and-after moment in this man's life mirrored the before-and-after moment in my neighborhood, and in the wider world.

It's obvious now that we made a grave system error. We sold the farm in the name of efficiency and "progress." We siphoned off the social glue of our neighborhoods. We sold off all our little stores, our lifeblood, our sense of community, in exchange for cheap prices.

What a trade-off.

To this day, economists have yet to calculate the social and psychological costs of a Walmart coming to town, of Alphabet owning Internet searches, of Meta commanding people-to-people communication and Amazon monopolizing online commerce.

So let's undo this catastrophic blunder. In addition to The 25% Rule, let's also impose A Size Tax on corporations — a progressive levy they must pay once they grow beyond a valuation that We the People decide is the max.

With this one, simple stroke capitalism will be upended. Small will be beautiful again. Neighborhood bakers, grocers, and hardware stores will spring back to life. Cities will churn with new energy. The internet will be a more diverse and fun place to go to.

But classical economists and business leaders will be horrified.

They will denounce the Size Tax as a crazy, bass-ackwards idea not even worth considering. So is turning in the direction of a skid — but it works.

AN END TO SHAREHOLDER PRIMACY

The deal for investors used to be: Enjoy your company's profits, but if the firm makes mistakes and causes harm, the cleanup is on you. As part-owner, you are responsible. That's the way it worked — risk it for the biscuit — until the mid-19th century.

Then came a landmark legal decision that shielded shareholders from personal liability for corporate missteps. This was rocket fuel for stock markets. Suddenly investors could shoot for the moon without worrying about future fallout. If a tanker runs aground or a chlorine plant blows up, or an ecosystem crashes, or rising seas start washing away cities, it's not your problem.

For everyone else and the planet, shareholder immunity has been a disaster.

Why don't we break this racket up?

Let's rewrite the laws so that shareholders are liable again. Oil spill? Poison in the food chain? Opioid epidemic? Lying about carbon emissions? Get your check book out, it's time to settle the accounts.

Financial markets, and indeed our whole business culture, will heave. Investors will immediately drop toxic stocks. They'll put their money into companies with healthier, more ethical cultures. Ones with strong environmental track records, no human-rights abuses, or rumors of executive misconduct.

Shareholders will be grounded.

Of all the memes we're proposing, this one might be the hardest sell. I can hear the pushback already: You freakin' idiots — Investors will vanish. Stocks tumble. Commerce grind to a halt!

Shareholder immunity is so baked into our current business model people can't even imagine it disappearing.

Well, imagine it.

A NEW CORPORATE DNA

These days, when a group of directors decides to bring a corporation into being, they lawyer up, draft a Certificate of Incorporation, file it at the Attorney General's office in the state of their choice and wait for the approval that automatically follows. Some states offer special incentives. Delaware offers freedom from liability, protection from unwanted takeovers, and complete anonymity of ownership (especially for offshore companies). It

also funnels corporate disputes into a "Court of Chancery" where no juries are allowed. Today, nearly half of all US corporations are registered in Delaware, and 70% of the Fortune 500.

The process by which We the People grant the privilege of existence to corporations has become a formality, a rubber-stamp job, a farce.

Now let's turn it back to what it used to be: a *negotiation*.

We the People hold all the power. We can refuse to grant a charter, amend or remove clauses, and add any provisions, requirements, demands and restrictions we deem necessary.

We can include a legally binding Corporate Code of Conduct that spells out the basic norms and values that the corporation must adhere to.

We can add a Transparency Clause that requires the corporation to be open to the public in everything it does; an Ecological Clause that instructs the corporation to be ecologically responsible in everything it does; a Penalty Clause that obliges the corporation to take full responsibility and financial redress for any social, environmental, or financial harms it causes. And we can include a Review Clause that tells the corporation it will be subject to a compliance audit of its charter obligations every five years.

We the People can make it clear to corporations that their right to exist is a privilege that We the People grant.

You exist on our terms.

Take a look at Article III on page 1 Meta Platforms' Certificate of Incorporation: "The purpose of the corporation is to engage in any lawful act or activity for which corporations may be organized under the General Corporation Law of the State of Delaware."

That's not a mission statement: that's a blank check. It says, we can do whatever we please. What Meta covets most is its collection of data-gathering algorithms. The company's success depends on keeping its working machinery secret — not just from competitors

EXHIBIT A
META PLATFORMS, INC.
AMENDED & RESTATED CERTIFICATE OF INCORPORATION

ARTICLE I: NAME

The name of the corporation is Meta Platforms, Inc.

ARTICLE II: AGENT FOR SERVICE OF PROCESS

The address of the corporation's registered office in the State of Delaware is 251 Little Falls Drive, Wilmington, New Castle County, 19808. The name of the registered agent of the corporation at that address is Corporation Service Company.

ARTICLE III: PURPOSE

The purpose of the corporation is to engage in any lawful act or activity for which corporations may be organized under the General Corporation Law of the State of Delaware ("General Corporation Law").

ARTICLE IV: AUTHORIZED STOCK

1. Total Authorized.

The total number of shares of all classes of capital stock that the corporation has authority to issue is 9,241,000,000 shares, consisting of: 5,000,000,000 shares of Class A Common Stock, $0.000006 par value per share ("Class A Common Stock"), 4,141,000,000 shares of Class B Common Stock, $0.000006 par value per share ("Class B Common Stock" and together with the Class A Common Stock, the "Common Stock") and 100,000,000 shares of Preferred Stock, $0.000006 par value per share. The number of authorized shares of Class A Common Stock or Class B Common Stock may be increased or decreased (but not below the number of shares thereof then outstanding) by the affirmative vote of the holders of capital stock representing a majority of the voting power of all the then-outstanding shares of capital stock of the corporation entitled to vote thereon, irrespective of the provisions of Section 242(b)(2) of the General Corporation Law.

2. Designation of Additional Shares

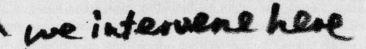

we intervene here

but from lawmakers and its own customers. It's a business model built on its clients' ignorance of how they are being used.

We can change this. We the People can decide that this corporation's m.o. does not serve the public interest. We can ask the Attorney General of the State of Delaware to add new terms and conditions to Meta's charter:

Consent Clause

Our company will unambiguously ask for permission to surveil and collect personal data. Our permission statement will be short and clearly written, so every person who decides to use our platform knows exactly what they are getting into.

Full Disclosure Clause

Our company's surveillance algorithms will be transparent and open to the public, so that every user knows exactly how their personal data is being collected, used and shared.

Now take a look at the "purpose" clause on page 1 of Goldman Sachs' Certificate of Incorporation. It's that same meaningless phrase that just about every corporation inserts into their charter. The one that gives it the legal right to carry out its business any way it wants with minimal responsibility to society.

We the people can decide that we do not like the way that Goldman has been behaving — specifically, its secretive and fraudulent behavior in the run up to the financial meltdown of 2008 and beyond. We can add a clause like this to Goldman's charter:

Financial Fraud Clause

If our company is caught manipulating stock prices, engaging in fraudulent trading, hiding information from regulators, or keeping vital public-interest data secret, then our business operations may be temporarily or permanently suspended, and our charter revoked.

CERTIFICATE OF INCORPORATION
OF
THE GOLDMAN SACHS GROUP, INC.

THE GOLDMAN SACHS GROUP, INC., a corporation organized and existing under the Delaware General Corporation Law (the "Corporation"), DOES HEREBY CERTIFY:

1. The name of the Corporation is The Goldman Sachs Group, Inc. The date of filing of its original certificate of incorporation with the Secretary of State of the State of Delaware was July 21, 1998.

2. This Restated Certificate of Incorporation restates and integrates and does not further amend the provisions of the certificate of incorporation of the Corporation as heretofore amended or supplemented. There is no discrepancy between the provisions of this Restated Certificate of Incorporation and the provisions of the certificate of incorporation of the Corporation as heretofore amended or supplemented. This Restated Certificate of Incorporation has been duly adopted in accordance with the provisions of Section 245 of the General Corporation Law of the State of Delaware. The text of the certificate of incorporation is hereby restated to read herein as set forth in full:

FIRST. The name of the Corporation is The Goldman Sachs Group, Inc.

SECOND. The address of the Corporation's registered office in the State of Delaware is Corporation Trust Center, 1209 Orange Street in the City of Wilmington, County of New Castle. The name of its registered agent at such address is The Corporation Trust Company.

THIRD. The purpose of the Corporation is to engage in any lawful act or activity for which corporations may be organized under the Delaware General Corporation Law. Without limiting the generality of the foregoing, the Corporation shall have all of the powers conferred on corporations by the Delaware General Corporation Law and other law, including the power and authority to make an initial charitable contribution (as defined in Section 170(c) of the Internal Revenue Code of 1986, as currently in effect or as the same may hereafter be amended) of up to an aggregate of $200,000,000 to one or more entities (the "Contribution"), and to make other charitable contributions from time to time thereafter, in such amounts, on such terms and conditions and for such purposes as may be lawful.

FOURTH. The total number of shares of all classes of stock which the Corporation shall have authority to issue is 4,350,000,000, of which 4,000,000,000 shares of the par value of $0.01 per share shall be a separate class designated as Common Stock, 200,000,000 shares of the par value of $0.01 per share shall be a separate class designated as Nonvoting Common Stock and 150,000,000 shares of the par value of $0.01 per share shall be a separate class designated as Preferred Stock.

add two clauses here

Now look at ExxonMobil's Certificate of Incorporation. Beneath all the legal jargon lurks the fact that this company is guilty of a crime against humanity.

Forty years ago, Exxon executives knew the implications of atmospheric carbon. But instead of coming clean, they sat around a boardroom table and hatched an obfuscation plan. They employed teams of "researchers" whose goal was to sow confusion about climate change, muddying the debate with lies. It was the most monumental corporate deception in history, and because of the time it wasted we may now be powerless to prevent the disruption, transformation and destruction of every city and ecosystem on the planet.

To stop corporations from inflicting this kind of harm on future generations ever again, we need a clause that is included de facto in the Certificates of Incorporation of *all* corporations:

The Kill Clause

If our company is found guilty of perpetrating massive social, ecological, or financial harm on society — in the form of repeatedly dumping toxic waste, damaging watersheds, fixing prices, defrauding employees or customers, or keeping vital information secret from the public — then our charter may be revoked, our assets sold off and the money funneled into a superfund for our victims.

In 1890, the highest court in New York State revoked the charter of the North River Sugar Refining Corporation with these words:

The judgment sought against the defendant is one of corporate death. The state, which created, asks us to destroy, and the penalty invoked represents the extreme rigor of the law. The life of a corporation is, indeed, less than that of the humblest citizen.

Imagine if We the People had won the battle with Philip Morris 30 years ago. Imagine if, instead of just slapping it with fines,

CERTIFICATE OF INCORPORATION
of
EXXON MOBIL CORPORATION

Exxon Mobil Corporation, a corporation organized and existing under the laws of the State of New Jersey, restates and integrates its Certificate of Incorporation, as heretofore restated and amended, to read in full as herein set forth:

FIRST. The name of the corporation is:

EXXON MOBIL CORPORATION

SECOND. The address of the corporation's registered office is 830 Bear Tavern Road, West Trenton, New Jersey 08628-1020. The name of the Corporation's registered agent at such address, upon whom process against the corporation may be served, is Corporation Service Company.

THIRD. The purposes for which the corporation is organized are to engage in any or all activities within the purposes for which corporations now or at any time hereafter may be organized under the New Jersey Business Corporation Act and under all amendments and supplements thereto, or any revision thereof or any statute enacted to take the place thereof, including but not limited to the following:

(1) To do all kinds of mining, manufacturing and trading business; transporting goods and merchandise by land or

add kill clause here

we sentenced it to death — had its charter revoked, its business disbanded, its brand erased, its profits paid out in reparations, its name tossed into the dustbin of history. If we made it an example of what happens to a killer corporation that betrays the public trust.

Hundreds of megacorps would be thinking twice before going rogue. Maybe General Motors executives would have opted for an immediate recall of its faulty ignition switch. Maybe Wells Fargo would have stopped short of signing its customers up with bogus accounts. Maybe the Sackler family would have balked on its opioid marketing scheme.

Maybe ExxonMobil would never have lied to us about climate change.

Democracy and rule of law and all the great ideals we believe in mean nothing when powerful corporations can commit heinous crimes and get away with just paying fines.

THE BREAKTHROUGH

Let's go after ExxonMobil with a vengeance and wipe it off the face of the Earth.

This will send a shiver down the spine of corporate America. Every predatory corporation that has betrayed the public trust, but seemed too big to fail, will suddenly be vulnerable. The relationship between people and corporations will heave. Capitalism will heave. The world will move in a beautiful new direction. And, out of this euphoric victory, a Corporate Charter Revocation Movement will be born.

This is how We the People get back our power, our agency, our sovereignty, our freedom and our democracy.

This is what the Third Force was born to do.

PETITION

To Revoke Exxon Mobil's Corporate Charter
IN THE STATE OF NEW JERSEY

Dear Attorney General of the State of New Jersey:

We, the undersigned citizens of the United States and New Jersey, who are sovereign over government and corporations, have the responsibility of keeping both of these institutions subservient.

According to New Jersey State law, you, the attorney general, may bring an action for the dissolution of a corporation upon one of the following grounds:

That the corporation has exceeded the authority conferred upon it by law, or has violated any provision of law whereby it has forfeited its character, or carried on, conducted, or transacted its business in a persistently fraudulent manner.

For over 50 years, Exxon Mobil Corporation has transacted its business in a fraudulent manner, persistently lying to the public about the the danger of carbon emissions, contributing to the heating of our planet, and thus perpetrating a crime against humanity.

Therefore, we the undersigned call upon you to commence proceedings to dissolve the corporate existence of Exxon Mobil Corporation.

April 9   2029

Gawd, we were so fucking stupid
... five mega corporations controlled
our eating ... three our drinking ... four
dominated our online lives ... five made
most of the world's cars ... two handled ride
sharing ... Big Pharma made the drugs ...

Now we're trying to reverse all that ...
make sure we control corporations,
not the other way around.

April 10

We've got rules now to keep corporations in check . . . and they're strictly enforced. Every industry must have at least half a dozen active players. No one is allowed to have more than 25% market share . . . and we have a powerful mechanism: a Progressive Size Tax (PST) for ensuring that happens. Basically, we tax corporations on the top of the heap until the desired level of diversity is achieved.

<u>Beats me why it took 200 years and a total system collapse to figure that one out.</u>

SECRETARY OF ▓▓▓▓

Thanks,

Why can't We the People Know everything?

On November 1, 1964, just as he was gaining real traction, Reverend Martin Luther King received a vicious blackmail letter. The anonymous writer threatened to destroy him personally and professionally, and suggested he just take his own life and save somebody a bullet.

Most now believe the letter was written by deputy FBI director William Sullivan, on the orders of J. Edgar Hoover. But no one saw that letter, or even knew it existed. The government's secret campaign to kneecap the civil-rights movement by taking out its leader only came to light after *The New York Times* unearthed the letter in 2014. Want to know more? You can't. All materials surrounding the case have been ordered sealed until 2027.

One of the biggest flaws at the heart of American democracy isn't the lack of a third political party that will do things differently, it isn't the apathy that keeps folks from voting, or fear of arrest that keeps them from protesting.

It's the secrecy.

We've let openness as a pillar of democracy fall away. From the highest levels of public life on down, an all-pervasive culture of concealment prevails. Every corner of government, from international trade pact negotiations to grand-jury deliberations, to the daily goings-on at the White House, the Pentagon, the CIA and the NSA is flat-out *infested* with an ethic of secrecy. Secrecy has become so commonplace that what should have been jaw-dropping geopolitical developments have passed by us, unnoticed.

In 2019 an American F-15 attack jet dropped a quarter-ton bomb on a crowd of people huddled on a riverbank in Baghuz, Syria. As survivors scrambled away, the jet dropped two one-ton bombs, finishing them all off. Seventy women and children were killed. Over the next two years the U.S. military, delayed, sanitized and classified every attempt to bring this war crime into the open. It only came to light two years later after a *New York Times* investigative team spent months piecing together leaked documents.

After the Taliban took over in Afghanistan, the *Washington Post* revealed that high-level government officials knew *eighteen years ago* that this war was becoming another hopeless, unwinnable quagmire, but they stayed the course and kept the messaging rosy.

This is nothing unusual. Most of the wars of the 20th and 21st centuries were hatched in secret.

The Vietnam war was triggered by a 1964 attack on a US destroyer in the Gulf of Tonkin that never happened. There were no enemy torpedoes, as announced — only an overzealous sonar operator chasing ghosts. But that soldier's erroneous report was the excuse President Johnson needed to persuade Congress to authorize the war. He knew the truth. Defense secretary Robert S. McNamara knew the truth. The only ones who didn't know the truth were the American people. And they would have no clue until Daniel Ellsberg leaked the Pentagon Papers seven years and a couple of million needless deaths later. After that a new verb entered the world. To be "McNamara'd" meant to be fooled.

Many of the dictators and strongmen of the past century were quietly propped up by the Pentagon and the CIA, and almost all the government-afflicted atrocities in Central and South America were carried out in the dark.

If World War III erupts, secrecy will almost certainly be the accelerant that ignites it.

How did we let this happen? Wholesale secrecy was never the plan for the United States. Just the opposite.

The framers knew secrecy has always been the irresistible tool of anyone in power, from kings and emperors and tribal leaders to corrupt schemers and petty rogues. So in drafting the constitution they made sure that checks on abuse of power were baked into the document.

One of the first things they did was enshrine protection for whistleblowers. "We, the people" would have an ear to the door. And for a century and a half any governmental attempts to draw the blinds always met fierce pushback.

But come World War II, when "national security" concerns trumped all else, a culture of concealment crept in. The newly minted adjective "classified" meant: "designated as officially secret: accessible only to authorized people." What kind of things were "classified"? Vital intelligence matters, military plans, weapons technology, the names of informants overseas.

But then the definition began to broaden. Routine bureaucratic business began to be classified. Even humdrum exchanges started being given one of the four labels that Truman created by executive order as the Cold War bit in: "Top Secret." "Secret," "Confidential" and "Restricted." There was always a reason to use one of those stamps, just to be safe. Government claimed the right to hold sensitive meetings in camera — and *everything* was arguably sensitive.

Even after the Cold War was over, secrecy remained a habit. The blinds stayed down.

And they're still down. Today, more than five million people in the US have the authority to "classify" information. Since the Nixon era it's become a reflex to classify anything remotely controversial.

When the Moynihan Commission into Government Secrecy looked into this in 1994, it found epidemic government overreach. More than 1.5 *billion* records, dating back 25 years, remained inaccessible. And if you get your hands on a classified document through the Freedom of Information Act, chances are it'll look like a blackout poem — heavily redacted, with entire paragraphs and pages smoked out.

In 2019, attorney general William Barr McNamara'd the American people by using secrecy, redaction and carefully orchestrated strategic delays in releasing Robert Mueller's report on Russian interference in the 2016 presidential election. The result was that most Americans thought that Trump had somehow been totally exonerated.

Imagine if Mueller's report had been released immediately in full for all to see (and why the fuck not?), then Trump's election, impeachment and his presidency may all have unfolded very differently.

Imagine if transparency, rather than secrecy, had been America's operating system from the get-go. If we'd stuck by the framers' guns and been bold enough to let the people know everything.

How might the progress of civil rights have been hastened had the government not been able to hide its tracks? How might America's political trajectory have changed had the government's covert counterintelligence program (COINTELPRO) — which targeted such dangerous subversives as feminists, peaceniks and the Black Power Movement — not been allowed to rumble undetected and unimpeded for a quarter century?

No Vietnam war.

No Iraq war.

No twenty years in Afghanistan.

Maybe no 9/11?

Maybe no Gaza genocide?

Maybe no WWIII?

Legitimate, high-level secrets — things like nuclear codes, the names of undercover agents, the details of hostage negotiations — are a tiny percentage of the material the government believes is too hot for us to handle. The bulk of it is mundane stuff that does not need to be concealed. We hear that Grand Jury deliberations have to be secret to shield the unjustly accused. Trade Deals have to be secret to keep their delicate socio-political engineering from being blown apart. Corruption investigations have to be secret to keep the targets from finding out they're in the crosshairs.

These are mostly red herrings.

The real reason so many public and corporate proceedings are classified isn't for safety or national security interests. It's to shield the actors and institutions from embarrassment and accountability. Secrecy is now mostly a way for people in power to cover their butts. To hide their dirty laundry and to push their agendas through with minimal resistance. It's the rotten smell that pervades our entire commercial, diplomatic and information systems.

I still remember the day, in 2010, when Julian Assange leaked those US army intelligence documents on Wikileaks. Thousands of pages of hot government secrets. And then, proving he was not to be cowed by cease-and-desist threats, he posted more incriminating documents, including Baghdad and Afghanistan war logs and videos.

Wikileaks suddenly introduced the possibility of *meaningful accountability*.

I thought: *Now* we're getting somewhere. Now our political system will be shaken to its core.

But it wasn't.

Three years later, Edward Snowden, a contractor working for the National Security Agency, encountered what he believed to be unconstitutional surveillance of American citizens, and he too went public.

This is why the framers inserted whistleblower protection into the constitution. The whistleblower is the foot soldier of democracy. It's only through the work of inside informants that we come to know what we need to know.

But our system cannot tolerate that. When whistleblowers stick their necks out, they are vehemently hunted down. There's every chance you get chased to Russia, never to be seen again, a la Snowden — or crucified, like Assange who got out of jail by pleading guilty to a felony charge under the US Espionage Act. A plea deal because he told the truth.

Every authoritarian leader's currency is secrecy and information control. They all try to stay in power by controlling the way information flows.

Vladimir Putin's invasion of Ukraine would have been unthinkable without his total control of the media. People believe his version of what's going on because that's the only version allowed on television and most social media.

The Chinese government is the most secretive of them all. It systematically blocks any online discussion that threatens its dominance. When a Politburo leader is accused of misconduct, all mention of it is immediately scrubbed out of view. When an artist, intellectual or activist asks for more freedom and democracy, they are swiftly removed and persecuted.

In America the populace is divided into info-tribes. Most people live inside their own confirmation bias. Few know what the truth is. Who could have imagined that tens of millions of Americans could be persuaded that climate change is a hoax, vaccines don't matter, and that Israel is only "defending itself" in Gaza. And with a new reign of Trump in America, the worst may be still to come.

THE END OF SECRECY AS WE KNOW IT

Secrecy is the elephant in the room of democracy. It undermines everything, and we'll never make it through the 21st century unless we change the way we *feel* about it.

No investigation, no grand jury, no trade negotiation, no cabinet meeting, no CIA strategic report, no confab between world leaders, no UN Security Council caucus can be so precious, so private, so sacred that We the People cannot be privy to what's going on.

This is a meme of disastrous consequence. We must feel outraged and fight back every time it happens. Make it the rallying cry of a new movement — Secrecy Activism, whose mission is to #MakeSecrecyTaboo in every aspect of democratic governance.

Here's how it could unfold:

We reframe the way we talk about secrecy. We start thinking about it the way we think about germs, viruses, bribery or insider trading — as something inherently filthy, something that needs to be excised from the body politic for our democracy to thrive.

Armed with this new vocabulary, this new *sensibility*, we embark on an all-out fight to purge secrecy at every level of governance.

No more hush hush goings-on in city hall. We remind our local politicians and civic leaders that they are not dealing with nuclear codes or high-level security matters. There is nothing, absolutely nothing, going on in City Hall that is so precious, so sensitive, that it cannot be shared with the citizens of the city. Total transparency from now on please!

No more police videos hidden from the public. We ask our police to keep their cameras on and tell our mayors that critical videos must be released pronto.

No more trade deals negotiated behind closed doors. We tell our policymakers that we would like to hear all the hot rhetoric and gamesmanship. Broadcast it on CPAC! Who knows, We the People may even decide to weigh in if the negotiations reach an impasse.

We tell our Presidents and Prime Ministers that we would like to listen in when they meet. Nothing they say can possibly be so sensitive, so embarrassing that We the People should not hear it.

We let the brass at the Pentagon and CIA know that We want to be in on pretty well everything that goes on around matters of war and peace. Try to cover up possible war crimes or keep vital information secret, and we'll jump on you with in ways you cannot yet imagine.

We let the Secretary-General of the United Nations know that the people's business should never be conducted in secret. Matters of global consequence at the Security Council must be debated openly so the whole world knows what's going on.

We jump on secrecy whenever and wherever it rears its ugly head. And when we fight, we're not just fighting to have some bit of information released . . . we're fighting for our dignity, our sovereignty, the inalienable right that We the People have to know everything and be the final arbiters of everything that goes on.

Imagine, We the People become *participants*.

What a concept.

When people feel empowered, they behave differently. The psychological power balance between the top and bottom of society is shattered.

Every major social movement of the last century is at least partly an attempt to expose a rot born of secrecy. People suffer in silence because powerful people are covering something up. #MeToo was the explosion of an age-old problem that had been contained by concealment. And what is #BlackLivesMatter but a reaction to the open "secret" of institutionalized racism — the mass denial of a grossly rigged game?

In the long run, the free flow of information is the only thing that works. It works at the level of nations for the same reason that it works for individuals. It's a good idea to come clean with your lawyer or your psychiatrist or your spouse. If you don't, your secrets will poison you.

Democracy moves in baby steps. First white men got the vote. Then Black men got the vote. Then women got the vote. And now everyone gets to know everything.

> An amendment to the constitutions of all nations and Article 19 of the Universal Declaration of Human Rights
>
> **Everyone has the right to live in a world without secrets. This right includes freedom from state deception, freedom from corporate misinformation, freedom from financial manipulation and the right to full public disclosure on all matters pertaining to peace, security, ecology and finance.**

That is how democracy evolves, how trust between the government and the people is restored.

Unconditional transparency can become an indispensable part of the mythology of the 21st century.

The stakes couldn't be higher. As long as elites and powerful forces are able to concoct wars and geopolitics in secret, We the People will never see a single day of peace on Earth. Not even a single minute. Aggression, hatred, greed, jealousy and fear may be the ingredients of war, but secrecy is the heat it needs to rise.

As long as secrecy in governance prevails, peace, unity and solidarity will always remain dreams.

On a personal level, there are skills we need to develop here. And they're not just reactive skills. Because secrecy isn't like a river you can see being dammed. It's rarely as obvious as some official avoiding your question, or failing to return your call, or holding up your access-to-information request. We need to train ourselves to detect *when information that should be available just isn't*. Like, when there is silence when there should be an explanation. Or when there is "noise" instead of sound — "noise" being some official policy statement full of weasel words so vague that all manner of shenanigans can go down in the shade of them. ("The president is authorized to use any and all appropriate force against those he determined planned or aided the terrorist attacks.") That's secrecy.

The work here boils down to training yourself to feel entitled to know what's going on.

Nov 10   2030

Secrecy was the great hidden monster that did us in. It was routine, all-pervasive, the rotten smell that waffed into every corner of political life . . . people never really understood what was going on . . .

There's a feeling now that secrecy is inherently filthy . . . something that must be totally eradicated for democracy to work . . . that the best way to avoid World War III is to have no classified information, no top secret documents, no redaction, no secret channels, no deep state . . . nothing kept hidden from We the People.

A fundamental shift in the nature of value

I've never quite understood finance. It's a nut I just can't crack. I don't get why the stock market goes up when there's bad news.

Or why at a time of climate crisis the Dow Jones is breaking all records. I don't know why 30 percent of new wealth is speculative — no physical objects bought or sold. Or exactly what work the three trillion dollars sloshing around the global economy every day is actually doing. Nobody has ever been able to explain this to me in a way that makes sense.

And how come tax havens still exist? And why did no one on Wall Street go to jail after the meltdown of 2008?

It's all a mystery to me.

But I think maybe I've been looking in the wrong place. Maybe the answer isn't in *The New York Times, The Guardian*, the CBS evening news or the books I've been reading.

My new theory is that a long time ago, capitalism got infected with a virus. And that virus seized control of its host and began to deform it. Markets drifted from their origins in honest trade between two parties, with both walking away satisfied. The purity of that transaction gradually got eaten away, until now, a millennium later, we have . . . whatever it is we have today. Junk capitalism. Predatory capitalism. Or simply a "death cult," as Jia Tolentino called it.

If this really is a doomsday sect then it all starts to make sense, this place we ended up. With the dehumanization of personal exchange. The explosive tension between rich and poor. The constant threat of speculative bubbles and ecological tipping points.

And maybe the virus, the actual virus, the one that has taken root in the heart of the global economy is ...

Money itself.

This thing that has grown and grown until it casts an immense cultural shadow, eclipsing security and trust and freedom and even love.

The story of this infection begins the moment we replace gift-giving and straight-on barter with some symbolic object: a shell, a coin, a paper note. This early money is a harmless, necessary thing — a placeholder for my chicken and your promise to build me a chicken coop. A portable bit of value that says, "Yes, I'm good for it ..." It's a leap of faith, really. We agree to believe in the value of the things being exchanged. We look each other in the eye and say *yes*, together.

But the thing about "trust me" is that there are no guarantees. Our best intentions are vulnerable to outside interference. This is what the old myths were about, how we puny humans are powerless when the gods start moving the furniture around.

And so as more and more distance creeps in, as money becomes more abstract, the virus mutates into ever more deadly strains.

In his infinitely insightful book *The Structures of Everyday Life: Civilization and Capitalism 15th–18th Century*, historian Fernand Braudel tracks the gradual insinuation of the money economy into the lives of medieval peasants. "What did it actually bring? Sharp variations in prices of essential foodstuffs. Incomprehensible relationships in which man no longer recognized either himself, his customs or his ancient values." His work became a commodity. And he himself became a "thing."

Because, see, as soon as there's money, lo and behold someone shows up whose job it is to . . . handle it. To keep track of our affairs for us. And naturally they need to make a living — just a little off the top.

So now there's this third party in the mix, between you and me, between my chicken and your promise to build me a chicken coop. A middleman. He isn't part of the actual exchange, but he's taking a cut of our trade.

And holy hell, that is the job to have. Because unlike our original transaction, there are no limits to it. Making money off of money is pure magic. That guy can get ahead no matter which way the wind blows. It's a sweet deal, because now there's real wealth to be had — it's in the currency itself — silver coins, gold ingots, printed notes.

A purity of intention used to run through the village: we are neighbors; we are trading partners. We are tight. What can I give you, brother? But now things have changed. You could be ripped off. So you look at that transaction with your neighbor with a touch of suspicion. Human interactions are no longer holy but transactional — every move has become a cost-benefit calculation.

Alienation creeps in.

The Catholic Church has a lot to answer for in all this. In 1095, it made a move no one saw coming: it started selling papal indulgences. This changed the very nature of money. Money was now this thing so powerful you could buy your way into heaven with it.

What a great business model for the church. The exploitation of medieval fears of damnation proved so successful that the sale of indulgences became an industry. Usury. Debt and the erasure of debt, a gateway to unfathomable riches.

Let the colonial pillaging begin!

When the Spanish conquistadores landed in the Caribbean and put slaves to work extracting gold, some mothers opted to kill their children and themselves rather than die slowly in the mines. For those Spanish sailors, looting and murdering with a view to earning more and more and more, the drive wasn't greed so much as shame and desperation. Everyone was caught in a debt trap. Living to pay off high-interest loans for their supplies and excursions. They could never get out of the hole. They sold their souls trying.

Usury was the original sin. Something horrible happened at that moment, in the West, when we started going down that path.

But this isn't really a story about money. It's about the choices we made to allow money to shape us. The moment money started using us, rather than the other way around.

So when was the tipping point? When did money stop being a tool and become . . . a craving?

Was it in the Middle Ages, when the goldsmiths who were storing folks' savings in their vaults realized they could crib a bit and issue receipts for money they didn't actually have in the vaults? That was the moment deceit got baked into the banking system.

Or was it in 1971, when President Nixon abandoned the gold standard — unhitching global currency from any real value at all?

Or was it 2008, when, as the banking crisis threatened to hobble the world economy, the governments of industrial nations used three trillion dollars of taxpayer money to bail out private financial institutions?

At what point did it become clear that money is simply a fiction?

And that those in charge of the storytelling hold the real power in a culture?

In truth there was no single moment that money became the lifeblood of our civilization, any more than there is a single identifiable moment when a virus starts making you sick.

And so here we are, in the endgame of financialized speculation, of digital hyper abstraction.

This is the most toxic strain of the virus.

It's out there, moving faster than the human mind, in volumes we can no longer comprehend: complicated bets made with borrowed money by computer algorithms. Financial assets like derivatives and credit-default swaps — the same instruments that stole the show in the 2008 meltdown and provide much of the global economic muscle now — bundled together and hawked as the world's best way to get rich. Speculations on speculations, hedges on hedges . . . these trillions coursing through the circulatory system of the world, this isn't real money, or even the idea of money. It is the idea of the idea of the idea of money. A consensual hallucination that rips through every human being on the planet with untold brutality.

The value of over-the-counter derivatives contracts now dwarfs the entire world's GDP. Debt has lost all meaning. The US has no intention of ever paying back the $32-trillion it owes.

A fundamental shift in the *nature of value* has taken place right before our eyes.

It didn't have to be like this.

What if we could walk this story back, as far as it takes, and then upend some key moments, so that a different story unfolds?

Maybe the rewrite begins in 1095, but this time the Catholic Church *thinks* about selling indulgences and then decides against it: *No!* And so the next step never happens, the step of virtuous "gain" — the idea that justified pillaging and slavery and consuming more than you produce simply by raiding the Colonial pantry — never happens. That idea is stillborn. Indeed, the church doubles down on the other side and prohibits the hoarding of treasure on the backs of others. With that moral weight against it, usury never becomes the bedrock of global finance.

And another world emerges.

What would be different now? Would we still have derivatives? Flash trading? Predatory loans? Perpetual growth? Or might those artifacts of capitalist overreach have starved in the shade of more humane ideals?

Maybe in this alternate universe our Western economic system retains a bit of the pure exchange ethic, worth for worth, the dignity and humanity of the exchangers intact. And maybe instead of a transactional system slowed at every turn by intermediaries reaching in to take their cut, we instead create a frictionless money system. Something a bit like the Islamic model.

Islam never went the usury route. It never took the bait. Islam said: We will not allow money to be made off money, we will not allow the rich to become richer just because they are rich. And with that decision they steered clear of the worst of the class warfare we're seeing now, the raging hatred that threatens to bring our whole Western house of cards crashing down.

The point is, we had a choice.

We still do.

The way forward is to kill off flash-bot trading, outlaw derivatives and credit default swaps, curb Wall Street's predatory appetites and flatten the money curve until the virus slowly dies.

That is our charge.

We start building towards frictionless money flows. We turn money into a public good, like highways. You can drive from New York to San Francisco without paying any tolls — so why can't you send $100 to your mother in Lebanon without someone lopping $5 off the top?

Cutting the middlemen completely out of the equation: that is the dream, that is our salvation. It's a long shot, a wild card, an almost impossible strategy to implement any time soon.

But as our current world system implodes along with our social and ecosystems, and market crashes start coming at us hard and fast, it might turn out to be the only viable financial play we have left in the planetary endgame.

Digital-financial hyper-abstraction is liquidating the living body of the planet and the social body.

Only the social force of the general intellect can reset the machine.

— Franco "Bifo" Berardi

The creative destruction of neoclassical economics

On September 15, 2008, out of the blue sky, came the crash.

Twenty percent of global trade vanished into thin air. The beginning of a depression that would last longer than the Great Depression.

Mainstream economists were blindsided. Not even one in a hundred saw it coming.

"*How did economists get it so wrong?*" asked the *New York Times*.

"*What good are economists anyway?*" quipped *Business Week*.

"*Will economists escape a whipping?*" wondered *The Atlantic*.

No "scientific" discipline has ever suffered such a blow to its credibility. Far worse than failing to predict the crash, mainstream economists actually enabled it, producing hundreds of reports supporting the dangerously leveraged mortgages and overconsumption that triggered the meltdown.

In London, one presenter at a meeting of leading economists stood and said: "We all got it largely wrong and have been using the wrong intellectual apparatus."

Student protests flared.

In the UK, a group of Cambridge students — spiritual descendants of the École Normale 15 — waved a manifesto called "Opening Up Economics."

University of Manchester students founded the "Post-Crash Economics Society." They published a book called *The Econocracy: The Perils of Leaving Economics to the Experts*, and a second called simply, *Rethinking Economics*.

The revolt spread to Harvard — home to the best economics department in the world if faculty Nobel-Prize-count is your metric. A hundred students walked out of Gregory Mankiw's introductory economics class. Eight hundred signed a petition to get an alternative to econ 101 into the curriculum. Over a hundred *Rethinking Economics* groups sprung up in universities all over the world.

At the 2015 American Economics Association (AEA) conference, delegates were greeted by a barrage of Adbusters' posters pinned up all over the corridors and meeting rooms of the Boston Sheraton. As night fell, we beamed provocations onto the hotel's façade.

The delegates could feel the breath of the resistance. *We're coming for you!*

This was a much bigger deal than the first round of student uprisings. The post-autistic movement had birthed a couple of dozen resistance groups. This time, more than a hundred mini uprisings spun out, in 70 countries.

But it didn't last long. The fury subsided. The revolutionary tumult of Occupy Wall Street came and went and not a single Wall Street exec ended up in jail. Not a single prof lost tenure. Not a single university changed its curriculum. The 24/7 news cycle moved on. A few passionate disruptors kept rattling the cage, but that was it. Once again, the keepers of the old paradigm held the line.

HOW SCIENTIFIC REVOLUTIONS REALLY HAPPEN

A theory, a paradigm, that has worked quite well for many years suddenly becomes problematic. Contradictions emerge; the theory no longer seems to predict reality. The scientific community huddles up. A flurry of experiments are conducted, information shared, papers written and conferences held. Out of this intellectual turmoil, a promising new theory emerges. It's subjected to rigorous scrutiny and tested in a myriad of ways. If it bears up, it's finally accepted as the new theoretical framework, the new norm, the new "truth." The scientists behind it are nominated for the Nobel Prize. The community settles back down, but now with a greater understanding of how the world really works.

This is a myth!

Thomas Kuhn, in his *The Structure of Scientific Revolutions*, tells us a different story of how paradigm shifts *really* happen. They are almost always nasty, messy, dirty affairs — very much like political revolutions, like putsches. The old guard jealously protects its turf. The dissenters are ignored, stonewalled, refused publication and tenure, ostracized and obstructed in every way — until tensions reach a boiling point.

This is the lesson we who seek to overthrow the neoclassical paradigm must learn: An old paradigm will not be replaced by evidence, facts, or "the truth." It will not be thrown out simply because its forecasts are wrong, its policies no longer work, or its theories are proved unscientific. An old paradigm will only be replaced by a new one when a group of fired-up mavericks orchestrate a coup, grab the old-school practitioners by the scruffs of their necks and toss them out of power.

A provocation projected onto the American Economic Association conference building, 2015.

"Is Economic Progress Killing the Planet?" poster at the University of British Columbia

THE COMING STUDENT REVOLT

"Externalities."

That's how generations of neoclassical economists have waved away pollution, soil degradation, biodiversity loss and ocean acidification. *Just the trims and ends of a rationally functioning market. No harm done.*

This approach has come back to crush us. Fifty years of dismissing environmental costs is more than a monumental blunder: it's flagrant professional misconduct.

"It's not even just about the right way to do economics anymore," says Australian economist Steve Keen. "It's about the survival of human civilization. If we are to have a future, then neoclassical economics has to go, and we heterodox economists have to replace it with something properly grounded in the physical reality of planet Earth."

At critical points throughout history, university students have catalyzed massive protests, called their professors and leaders on their lies and steered their nations in brave new directions. We're at another of those do-or-die moments now.

Students today are called upon to seize the moment and help engineer the overhaul of a whole toxic economic ideology that has led humanity to a dead end.

A provocation at the University of British Columbia

The revolt could unfold something like this:

Activist cells spring up on campuses around the world. They disrupt classes, post provocations in the corridors and confront their professors in increasingly angry ways.

Every Friday they nail *KickItOver* manifestos to their professors' doors. They use big medieval nails, a la Martin Luther at Wittenberg, and post videos of their raucous ceremonies on the Internet.

A ritual is born.

And that is the beginning of the Economic Reformation.

Books come out, lots of them — not just critiques in the mold of Piketty and Duflo but passionate calls for the overthrow of neoclassical orthodoxy and its toxic marriage to the war economy at the expense of humanity and the planet.

Place strategically all over campus

A dark-horse maverick wins the Nobel Prize for economics.

Another financial meltdown hits. This time there's no next-day correction by bargain-hunters jumping in. Instead, the markets just keep tumbling: New York, Paris, Frankfurt, Tokyo.

The #KickItOver Manifesto becomes a rallying cry for a new kind of economics. Students burn textbooks, occupy their departments, issue ultimatums. IMF and World Bank HQs are invaded, stock exchanges ransacked, the Federal Reserve swarmed. Finance ministers are pied and pelted with rotten eggs wherever they go.

And the broader public goes, Yeah, you know what, those dinosaurs had it coming. It's time.

Five hundred years ago Copernicus proposed a new model of the universe — and suddenly the stubborn math that had befuddled the Ptolemaic astronomers magically fell into place; if the earth circles the sun, rather than the reverse, it works!

The science of economics is in an eerily similar situation right now. How could we have missed this simple truth? That the human money economy is a subset of the Earth's larger bioeconomy, not the other way around.

This shift in perspective changes everything. It invites us to see the world with fresh eyes.

Feminist economist Lourdes Beneria exposes neoclassical economics as wildly overfocused on men. The "household economics" models are stuck in a sexist, 50s-era world lacking nuance or understanding of gender relations and the "bargaining" that goes on in ways econometrics can't capture. The bloodless rational models don't reflect how much emotion is behind how decisions are actually made.

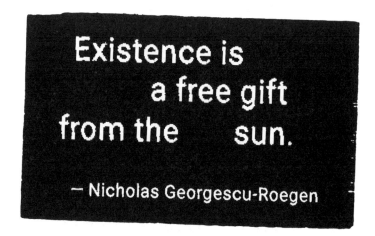

A COPERNICAN SHIFT IN

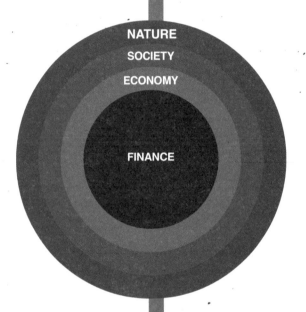

ECONOMICS

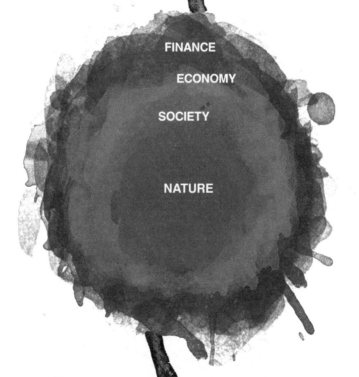

Who grows up happier:
The suburban kids of North
America, or the kids in the
slums of Dhaka?

Poverty economist Manfred Max-Neef sees the current language of economics as the language of the elites. It means absolutely nothing to people who can't afford shoes. In fact, it's an affront, it's absurd. Our current crop of economists have little insight into the behavior of 80 percent of humanity. They haven't got a clue about how to lift the world's poorest out of their misery.

A few years ago, Hazel Henderson — the maverick thinker who beguiled me back in the '80s — published a manifesto called *Statement on Transforming Finance Based on Life's Principles*. It's based on the principles of ethical biomimicry hatched by biologist Janine Benyus.

"The economy of the future," she writes, "will be based on "strategies universal to all organisms." It is a set of instructions based on nature's 3.8 billion years of successful learning and innovation. Henderson and Benyus call these instructions "life's best practices." And we too must follow them as we build social and financial systems and design our communities going forward. The bedrock of biomimicry economics is to change the way we think about growth. It's not wrong to think about optimizing growth, but this is not about the growth of one species, but the growth *of the whole planetary biome*.

In 2022, *Adbusters* talked to Herman Daly, the father of environmental economics, in what turned out to be his last interview before he died of a brain bleed at age 84.

For much of his life, Daly was a prophet without honor in his own discipline. He was widely vilified for questioning the gospel of growth. Colleagues trash-talked him behind his back. His students had trouble getting jobs. Only in the last few years was a new generation taking up Daly's refrain that we have to think of economics differently now. Was Daly pessimistic? Nah. "It's too late to be pessimistic," he said. The sad irony is that it took the five-alarm

KickItOver Manifesto

We, the undersigned, make this accusation: that you, the teachers of neoclassical economics and the students that you graduate, have perpetuated a gigantic fraud upon the world.

You claim to work in a pure science of formula and law, but yours is a social science, with all the fragility and uncertainty that this entails. We accuse you of pretending to be what you are not.

You hide in your offices, protected by your jargon, while in the real world forests vanish, species perish, human lives are ruined and lost. We accuse you of gross negligence in the management of our planetary household.

You have known since its inception that your measure of economic progress, the Gross Domestic Product, is fundamentally flawed and incomplete, and yet you have allowed it to become a global standard, reported day by day in every form of media. We accuse you of recklessly supporting the illusion of progress at the expense of human and environmental health.

You have done great harm, but your time is coming to its close. The revolution of economics has begun, as hopeful and determined as any in our history. We will have our clash of paradigms, we will have our moment of truth, and out of each will come a new economics – open, holistic, human scale.

On campus after campus, we will chase you old goats out of power. Then, in the months and years that follow, we will begin the work of reprogramming the doomsday machine.

Nail (Martin Luther style) to professor's door

fire of the climate emergency for the skeptics to weaken their resolve against him.

"Big changes usually require a big crisis to make them politically possible," Daly told us. "Maybe things have finally become sufficiently dire."

The economists of the future will step down from their towers of perfect rationality, draw inspiration from psychology, sociology, anthropology, mythology and neuroscience, and start taking some baby steps into new worlds of emotion, social relations, religion, morality and values. They will do rigorous fieldwork in the streets, slums and favelas of the world. They will know what it feels like to go hungry, to lose your home, to have to resort to crime to pay for a life-saving drug. And they will come to a stunning conclusion: When the chips are down, cooperation rather than selfish interest is the key to survival.

The successful economies of the future will be less about interest rates, tax codes and money supplies and more about innocence, playfulness, empathy, trust, social organization, and figuring out the real cost of things.

A new breed of *bioeconomists* will abandon the dream of creating a rigorous, math-based science on the model of physics. Instead they will nurture a new dream: to create a *bionomics* — a social science, that cuts across and unifies all the other sciences into one all embracing super-science.

A few years ago when I gave a talk at the University of British Columbia, none of the professors even bothered to show up. They are a cowardly bunch of logic freaks who hide away in their offices while the planet burns.

To shift the economic paradigm and win the planetary endgame, students around the world will have to dig deep and find within themselves a new kind of fury.

Dec 3, 2031

Maverick economists everywhere are on a rampage . . . they're determined to kill off all the remaining "logic freaks" . . . come up with real measures of progress . . . make all markets true-cost . . . outlaw derivatives . . . abolish flash trading . . . ultimately make usury taboo.

Occupy Finance!

We're now arrived at the existential crisis I've been dreading for the last twenty years. A perfect storm of doomsday scenarios — the grinding anxiety of our precarious jobs and our precarious lives, the explosive tension of the rich-poor gap, the constant threat of speculative bubbles and ecological tipping points, mired in wars that threaten to tip into WWlll at any moment, and on top of it all, the helpless feeling of being trapped in a system of surveillance that won't let go of you no matter what you do — all this has seeded a level of desperation beyond anything we could ever have imagined a decade ago in Zuccotti Park.

Big Finance is in many ways at the bottom of all of it.

I wake up every morning half-expecting to find the Dow Jones has plummeted ten thousand points and we're in a worldwide meltdown that will never end. The monolithic structure of global finance is its own kind of tyranny, and nothing good can happen on Planet Earth until we find a way to seriously rehabilitate it. No real progress can happen until we restore *value* to the heart of the exchange equation. Until we build a vibrant economy that rewards real innovation, one that's about making useful things and providing needed services — real work and real output, rather than the useless offgassing of high finance. Derivatives and credit default swaps and crypto leveraging: these things did not even exist 20 years ago. Now they drive a multi-trillion-dollar game that will take us down with it when it collapses.

#OccupyWallStreet made some noise and politicized a generation, but we didn't fix things. Now it's up to us — *all* of us. To win the planetary endgame, we must take on this lifesaving task. We must attack the soft underbelly of Big Finance and fight like hell for a new sense of value.

If we apply enough pressure where it counts, we'll start generating movement in the right direction, and pull off a kind of system-wide recalibration.

STEP 1: WE ELIMINATE ALL TAX HAVENS

A huge chunk of planetary wealth is simply unaccounted for. With the help of lawyers, accountants, tax consultants, and complicit governments, rich people are hiding it in tax havens. There's far more money in these OFCs (Offshore Financial Centers) now than there is actual currency in circulation: more than 30 trillion dollars.

And most of the global banking system is in on the scam. The Panama and Pandora Papers leaked in 2016 and 2021, implicated 500 banks, numerous heads of state, thousands of politicians and public officials and 130 billionaires all caught up in a global network of over 200,000 offshore accounts.

This is a rigged game played by Russian oligarchs, Middle Eastern inheritance-tax dodgers, arms dealers and tens of thousands of anonymous wealthy folks trying to become wealthier. "When it comes to the question of paying taxes," the economist Gabriel Zucman says, "the rich have seceded from the rest of humanity."

A crackdown on international evasion is difficult because it requires international co-ordination. But it can be done. Effective legal instruments exist to prevent offshore tax evasion. All we have to do is make it illegal for banks to enact transactions with territories that don't comply with rules on tax transparency. That one simple rule closes them down instantly.

The bottleneck is political will. Our gutless leaders, under pressure from lobbyists, will always avoid jumping. They've got to be pushed. We the People must send them an ultimatum and make them pay a heavy price if they don't comply.

STEP 2: WE CALM SPECULATIVE FERVOR

So you want to "get into the market" because that's where the real money is. Couple things to know: Most of the folks you'll be competing against aren't human — they're bots. Their nervous systems are algorithms that detect fleeting price discrepancies and market patterns, and then place orders automatically.

That's okay, right? Humans played chess against computers for the fun of it and even beat them until recently, right?

Except that the companies you're up against, the ones engaging in high-frequency trading (HFT), are playing a different game — one no investor *can* play. In theory you're at the poker table with them, but the bots are really playing each other and the edge they're chasing is measured in nanoseconds. So investing isn't about investing anymore. It's not about looking for promising companies and supporting them to the level of your risk tolerance. It's about hardware and software. It's an arms race: and the table's tilted toward the companies like Citadel that have the money to build the fastest uplink closest to the stock exchange — to push their trade through a nano-whisker ahead of everyone else. And in fact most observers think the arms race will soon become a space race, with big players jockeying to colonize lower-earth orbit with necklaces of hundreds or even thousands of satellites so that one will always be close enough to get their "buy" or "sell" signal in there first as they shoot money across a dozen global stock exchanges and unofficial "dark pools" of high-volume speculation — leaving earthbound punters with their lasers and fiberoptic cables in the dust.

About three trillion dollars a day is traded this way. It creates nothing of actual value. It's the ultimate financial circle jerk — a massive-scale transmogrification of money that means nothing. Except that whole thing is a ticking time bomb, and when it goes off, we're all going down.

There is a solution. It's simple but not easy.

It's a "hold" rule.

You put a legally mandated gap between the trade and its acceptance. So when someone buys a stock, they have to hang on to it for some period of time before they can sell it again. Let's say it's 24 hours. Now people start looking again at what it is they're buying. The actual value of the stock, rather than its role in the spasm chain of money-begets-money-begets-money. Nanotrading: gone. Insider trading: severely curtailed. Wall Street grounded. The intercontinental money rocketing between exchanges in North America, Europe and Asia: hobbled.

The 24-hour holding rule works. You buy it, you keep it for a day. Elegant, simple, radical.

We fight for this and we win, because the vast majority of the people in the world want it.

STEP 3: WE IMPLEMENT THE ROBIN HOOD TAX

The Robin Hood Tax has tantalized activists ever since Nobel-Prize-winning American economist James Tobin first suggested it in the 1970s. He thought of it as simply a charge on foreign-exchange transactions. But slapping a tax on all high-end stock-market plays — from stocks and bonds to derivatives — would cool speculative action in the sandbox of the wealthy, and redistribute that money where it can be a force for good.

The tax wouldn't have to be high. Even scalping 0.05 percent off the top of each foreign-exchange transaction would generate $100-billion a year. That could be used to all-but eradicate extreme poverty, provide universal health care, slow global warming and deal with unexpected global crises. Raising it to one percent would generate a whopping two trillion a year.

Sweden was the first to introduce this tax in the 1980s. But speculators simply moved their money to other markets. After seven years, Sweden had no choice but to give it up. The Swedish experiment made it clear that a tax on stock and currency trades will only work if it is introduced globally.

We have to do this together.

STEP 4: WE MAKE DAILY OFFERINGS TO THE MONEY GOD

There are places we go to stand, naked and vulnerable, before a higher power.

Like the ATM.

We make our pilgrimage to the money machine, say a little prayer for solvency, slide our card into the slot, and out of the machine comes a few bills into our waiting hand, like a wafer on the tongue. ATMs are constant reminders that capital is the dominant religion of our time. They are modern shrines.

So let's treat them that way.

Let's drape ATMs with garlands, decorate them with flowers, crucifixes, votive candles, incense burners, little statues of the Virgin of Guadaloupe. Let's surround them with mementos of all that we sacrifice in the struggle for economic survival: our time, our families, our freedom, our joy. Let's place at ATMs the photos of friends and loved ones who took their own lives because they could no longer live with unemployment, homelessness, or hopelessness. Let's decorate them with little postcards of Jesus throwing the moneylenders out of the temple, and with visions of a better and fairer world. Let's turn them into symbols of the fight for a new kind of meaning. And on Sundays and May Days and paydays let's gather around ATMs and dance and perform exorcisms and voodoo rituals.

We drape ATMs with garlands, flowers, crucifixes
and little statues of the Virgin Mary

What happens then after a while is that the ATM no longer belongs to the bank. It belongs to us. We the People. The children of a dark future and the next-generation architects of salvation.

STEP 5: WE FLATTEN THE MONEY CURVE

Covid-19 turned from an outbreak into a pandemic because of airplanes. The virus shot around the world, instantly found new hosts, and replicated everywhere all at once. It was out of Pandora's box before anyone could shut it down.

Speculative money has spread the virus of capitalism the same way. The throughput is out of control — because there are no borders or governors or effective laws to stop the money-begets-money-begets-money algorithm. The speculative traders are like superspreaders, and the whole world is vulnerable.

Our only hope is to shut these guys down. The new financial architecture must *flatten the money curve.*

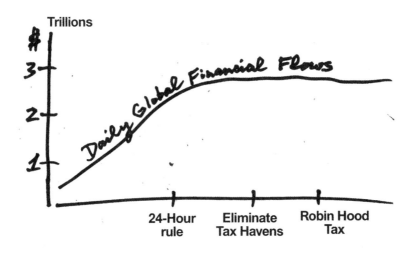

We make it harder and harder for speculators to ply their unholy trade. We reduce the money sloshing around the internet every day from $3-trillion to $2-trillion to $1-trillion . . . and then down to a

the GREAT GLOBAL POTBANG
for a new financial architecture

This May Day,
we bang our pots
and pans:

For the elimination of tax havens

For a 24-hour hold on all stock trades
(You buy it, you keep it for 24 hours)

For a 1 percent Robin Hood Tax on ALL stock market transactions and currency trades

modest $500-million. Gradually, the flow of toxic money subsides and the flow of money that's doing honest work grows, until we reach financial herd immunity.

You may think all this is a bit of an idealistic, romantic fools' game. Who could possibly take any of this seriously? But wait until the global temperature creeps up a bit more; until the heat domes and arctic blizzards become unbearable; until the next financial meltdown hits you hard and your life falls apart like it already has for hundreds of millions of people in dozens of failed states; wait until your bank won't let you in the door and your anger turns into panic — then you may suddenly be ready to take on the leeches and parasites of Wall Street and get behind some of the systemic heaves to the world's financial architecture described in this chapter.

h, oh my god . . . the bubble just popped . .

Nov. 7   2032

After the Crash, we
   ransacked Wall St. . . .
occupied stock exchanges
 . . . did some really bad things
to BlackRock, Goldman,
                JPMorgan,
   Bank of America, etc. . .

We brought Big Finance all the
         way down,
   made sure it would never
dominate our lives again.

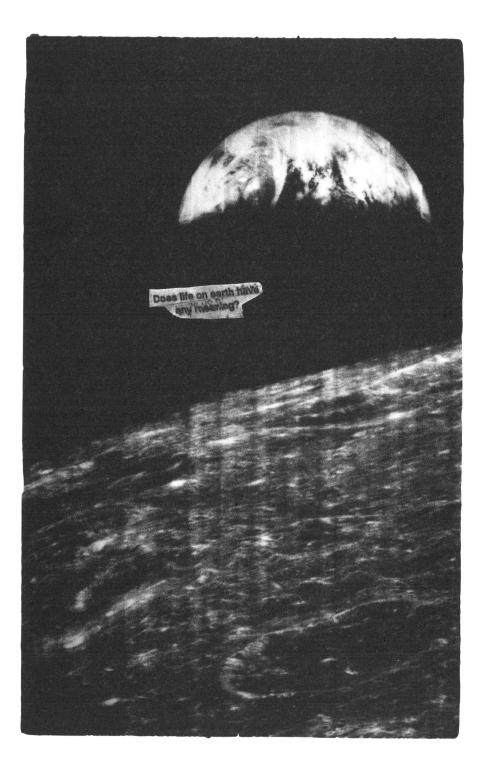

disillusionment

At the apex of the *Adbusters*' flight path, as our worldwide circulation climbed beyond 100,000 and our NGO was becoming a geopolitical force to be reckoned with, a strange thing happened. I became disillusioned. Or rather, I became profoundly disappointed with the other kids in the sandbox. I felt that all of us had lost our way.

We Lefties just weren't . . . making things happen. A protest could draw a million people, but then everyone went home and resumed their routines. Internet campaigns whipped up great anticipation, but a week later you hardly even remembered what they were about.

We whined a lot. Read many books. Knew what was wrong with the world but had no idea how to fix it.

When did we start taking ourselves so seriously? Where did the "people's laughter" go? Weren't we the ones who knew how to live, love, think and have fun? And, you know . . . care? About the big stuff?

I saw my compadres fighting smaller and smaller battles. Choking on critical theory. Eager to cancel anyone who disagreed. A sort of self-purification program seemed to be happening — an "I can hang my head lower than you" one-upmanship of gooey liberal guilt that was frankly embarrassing.

The Left suddenly just wasn't cool anymore.

And that's not trivial. Because when you lose your cool you lose your influence. Cool's sneaky superpower is its ability to propagate, to replicate in a critical mass of human minds. Lose that and you've lost the game. After 50 years of identifying as a Lefty, I realized I couldn't in good conscience wear that label anymore.

And then, at the peak of my disillusionment, the story took a new twist.

Young people rose up and chased corrupt President Ben Ali out of Tunisia. Tahir Square kept pulsating until Egypt's long-time dictator Hosni Mubarak fell. The Arab Spring spread quickly through the whole Middle East. It was exhilarating. In our brainstorming sessions we kept asking ourselves: Why can't we do something like this? Why not . . . an American Spring?

We got excited about the idea of marching into the iconic heart of global capitalism and occupying it.

We chose September 17 as the kickoff date because that was my mother's birthday. For me, it was personal.

My mother, Leida Lasn, was the soul of the image we chose: a ballerina on a bull. Serenity in chaos. Like a Basho poem read on the floor of a stock exchange, or welders at Toyota laying down their torches and praying to the fire gods.

We started pumping out tactical briefings.

The first one read:

```
Alright you 90,000 redeemers,
rebels and radicals out there,
```
On September 17, we want to see 20,000 people
flood into lower Manhattan, set up tents,
kitchens, and peaceful barricades, and occupy
Wall Street for a few months. Once there,
we shall incessantly repeat one simple demand
in a plurality of voices. This could be the
beginning of a whole new social dynamic
in America . . .

We started collaborating with activists on the ground in New York. David Graeber and local community organizers mustered. And on September 17 it happened: a peaceful battalion of a few thousand marched from their base camp in Zuccotti Park to the statue of the charging bull in the heart of the financial district.

From the wings came support over a ridiculously wide spectrum — from the Anonymous collective to Michael Moore to Elizabeth Warren to Barack Obama. Electric speakers, from the philosopher Slavoz Žižek to actress Susan Sarandon — kept occupiers' spirits high.

It was leaderless, joyous and intense.

The energy built, fanned unwittingly by Mayor Bloomberg's tone-deaf pushback and a NYPD beat cop caught on camera viciously pepper spraying a couple of teenage girls. On October 3, more than 700 protesters were arrested on the Brooklyn Bridge.

And it spun out. Occupations popped up in Chicago, LA, Seattle, Atlanta, Miami, Washington. The spark jumped borders: Toronto, Montreal, Vancouver. And oceans: London, Paris, Madrid. Sydney. Jakarta. Bangalore. By October, there were occupations in more than 2000 cities around the world.

This was more than a protest against greedy bankers and the shenanigans of the global financial system. It was a great howl from the soul from a whole generation that said: We are the 99% and we're sick of you 1% lording it over us.

It felt uncannily similar to what happened in 1968 when a small student uprising in the Latin Quarter of Paris spread like wildfire, across borders, across oceans. For a few heady weeks in 1968 and again in 2011, a tantalizing question hung in the air: Could this be the beginning of a world revolution?

Looking back, what killed the momentum in both 1968 and 2011 wasn't that people failed to put their asses on the line. They did. But we failed to deliver what every revolution needs: one big issue to galvanize around. We didn't have our memes figured out. We had the attention of the world but couldn't muster much beyond a few cryptic pronouncements. We kept shouting, "We are the 99%!" But when winter crept in and Mayor Bloomberg ambushed Zuccotti Park in the wee hours of the morning, we lost our momentum and will to fight back.

OWS fizzled out, went out with a whimper. At least that's what people say.

But I don't quite see it that way.

Those three months in 2011 reinvigorated the Left and stoked a debate about inequality that's still gathering momentum. You can trace a direct line from #OWS to the rise of #MeToo, #BlackLivesMatter, #FreePalestine and the protest movements now springing up around the world. Occupy Wall Street politicized a whole generation, just like my generation got politicized back in 1968.

One evening in 2019, while reading a *New York Review of Books* article called "Two Roads for the New French Right," I had an aha moment.

The cultural critic Mark Lilla was surveying the European political landscape. And what he saw excited him. Something fresh was coming up between the old intellectual swamp gas of the Left and the poisonous xenophobic populism of the Far Right: a generation of intellectually daring young politicos who aren't easily categorized.

They put out small magazines that punch way above their weight. ("The point of little magazines is to think big in them," as Lilla put it.) These magazines are peppered with references to George Orwell, Simone Weil, Hannah Arendt, the young Marx, the cultural conservative American historian Christopher Lasch . . . They argue about Heidegger, publish critiques of neoliberal economics and environmental policy as severe as anything you can find on the Left. They rail against unregulated global financial markets, neoliberal austerity, genetic modification, consumerism, soulless modernism.

They say that the European Union has been centralizing all the power in Brussels, conducting a slow coup d'état in the name of economic efficiency.

Some of them believe in zero growth. They want Big Tech reigned in — calling for a dismantling of the Google-Facebook-Amazon troika that's mindfucking us.

They're passionate. Committed. They have a plan.

The catch? There is a strong strain of social conservatism about them. The Left wants nothing to do with these people because their plan comes from an ideology that's grounded in tradition, and even faith.

That sounds radical to me. In the best way.

It speaks to a spiritual dimension that's gone AWOL among progressives. A dimension I fell in love with in Japan. So many of the Shinto-based rites I explored in my documentaries are about cherishing your ancestors and acting in a way that reflects honor upon them. Living lives our loved ones would be proud of. That attachment to the past, to tradition, to legacy, gives us meaning. These are stories that started long ago and will continue long after we're gone.

Lilla had somehow exposed the bigger picture that the Left, I feel, has turned its back on. Yes, this is about the 'good fight' for equality, human rights and justice, but it's also part of an "organic continuity" that stretches back thousands of years. On this view, the fundamental task of society is to pass on a kind of moral code to future generations — to put people into the larger organic flow of things.

What the job of society definitely isn't, is to become what Lilla calls a clubhouse for "autonomous individuals bearing rights."

When I read that, I thought to myself, Fuck, this rejection of hyper-individuality is what I've deep-down believed in all my life.

For the last 20 years, I've increasingly felt that we on the Left are marooned in a dry place and don't know how to make the flowers bloom. There was a piece of the puzzle of living a good life that the

Left wasn't giving me. And suddenly, I sensed it, felt it, unbelievably, coming from the Right.

One of those Lefty shibboleths — and I know I'm going to get into trouble here — is abortion.

Since *Roe* v. *Wade*, a woman's right to choose has become such a given that it is now a reflex. Somewhere the lofty principle of justice got uncoupled from the primal human response to the sacrifice itself. Because sitting with the triumph and the horror at the same time is just too hard.

To say that abortion makes you uncomfortable isn't the same as saying you're constitutionally against it. But I dare you to try it. The slightest hint of ambiguity around the issue strikes the Left as outright treason, and the blowback you get is calibrated to kill.

I find it strange that progressives can be "pro-life" in every context but this one. We fight tooth-and-nail against the death penalty, against war, against torture and state-sanctioned killings across the globe. Even meat is murder. Progressives are about the sanctity of life above all else except in this one domain.

And I get why this issue's different. Don't misunderstand, I've always felt that a woman has the final say about whether to go through with her pregnancy or not, and nobody should ever take that right away from her. All I'm saying is, if a life is taken, it should be taken with the same reverence as happens in every other context. Whatever a fetus is, it is not medical waste.

There's a Japanese Zen ritual called mizuko-kuyo — literally the "water baby" — that has found its way to America. As practiced in the States it is surprisingly politically generous. This is not just something pro-choice families embrace. It is a sacred rite. We mourn lost life here no less than the descendants of soldiers mourn the dead at cenotaphs, or Argentinian mothers mourn those "disappeared" in that country's Dirty War. An activist drugged and thrown from an airplane into the ocean; an unborn baby sectioned and vacuumed from a womb: if the thought of both of those doesn't make you hang your head a little, paralyzed with emotions bigger than you can hold, honestly, I'm not sure where we go from there. If we can't find a place to meet in things that are the very denominator of our species, what hope is there for us?

We Lefties have lost our way. We've become cold and egocentric. That's why we've been thrashing around in the weeds for the last twenty years. Now the time has come to take a leap into the unknown and learn to play jazz again.
This time with a new sense of awe.

Nov 13

Last night was chilly and there was
no moon . . . and out there on my
midnight walk with Taka-chan . . .
I started thinking about the
    whole geopolitical mess we're in
    — populism, authoritarianism,
    countries split down the middle,
    hordes of honest souls believing
    all kinds of crazy conspiracies,
fascism on the rise *again* . . .

FREENET

Nov 14

. . . now I remember . . . it was 2008 (ish), springtime (for sure) and I'm in New York for a conference. I'm lost and I'm late and I ask a guy who looks local: Any idea how I get to 323 William Street? And instead of replying, he pulls out his phone. In ten seconds he's got it. He shows me the screen. Google Street View swooping us around two corners and right up to the front door of my destination, holy shit.

Nothing we've ever done as a species compares with this.

Dec 29

My mind goes back to that unforgettable day when the first ads appeared on the internet. People were outraged. Like

no! no! not in this space, ever!
Fuck off, you can't have it!

But the fury died. People got used to it. Because this thing was clearly not going away. The commercial virus had found its perfect host. And now it was on a tear.

With TV at least you could turn it off.

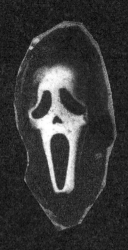

Feb 4

Once you make a system error
like this — mixing commercialism
with communication . . . accepting
surveillance algorithms as
the norm — it's hard to undo it
later, downstream. The error
compounds . . . it becomes
a river choked with logos
and memes and cookies and
avatars and push messages and
influencers, all bubbling along in
one perfect frictionless flow . . . a
greased slide deeper and deeper
into the Heart of Darkness.

Mar 25

We're in humanity's second great migration now.

The first one, a billion or so years ago, was the one from ocean to land. Evolution is slow but it's self-correcting and perfect ... a drama unfolding over millions of years, nature climbing the solution tree ... organisms pushing toward their full potential ... swim bladders evolving into lungs. Fins becoming limbs.
Eyes growing bigger. Till, boom! Our distant ancestors claimed this rock.

But just as we'd set up camp ... an extinction event wipes out most species.

Nature reboots.
The machinery starts working its crystalline perfection again ...

Only for yet another extinction event to upset the board all over again.

And another, and another ... Until finally the smooth apes emerge. And we learn to walk and grow food, build civilizations ... and fly to the moon.
Our first great migration turned out pretty well in the end ... But it wasn't the end — that much is now clear. We're on the move again. Only this time, we're migrating right out of our bodies ... and into a strange new kind of terrain.

Can we live in a totally human-constructed space? There are no wild beasts there. No running water . . . no mosquito bites . . . no skin to touch . . . no real sex to have. It's all just ones & zeros.

Can this really replace . . . everything else?

I'm not real.

April 17

Too many of us are wallowing in our virtual playpens while ecosystems crash outside. We say: "Ahh, I don't really care that much about the physical world, I'm happy here in my metaverse. Whatever this dream turns out to be — rocket to transcendence or evolutionary blind alley — I'm in." What could very well happen then over the next few decades, is that the mercury keeps rising, our planetary systems collapsing, and we're driven to live more and more in our private little digi-cocoons.

Until . . .

at some point we wake up and go, wait, what? No! No! Dammit, no! I want out of here . . . I want to go home! I want to eat a good burrito . . . make a baby . . . walk on the grass . . . bury my father . . . lock eyes with a wild moose . . .

But it will be too late. The natural world will be pretty much gone. We'll be deep into the next phase of human evolution . . . thrashing around on the electronic beach, destined to spend the next million years stewing in our own juices.

April 23

In China they've already accepted
total surveillance.

The terms are basically: I accept being
bird-dogged and kept in place by this
social-credit system that rewards me for
doing what the government wants.

And I'm pretty sure we're next... we'll roll
over like that too.
As our summers become insufferable,
financial meltdowns more frequent
and survival more precarious, people
will raise the white flag and say, OK, I'd
rather have a good meal and a decent
job and a roof over my head, and I
don't really care that much about freedom.
Go ahead! Install an app on my phone.
Track me. Harvest my banking info. Put
a hundred million cameras in the public
square. Capture my face and read my
blood pressure, so you can identify me as
a dissident when my blood boils during the
Great Leader's speech.
In nation after nation, we'll
accept massive surveillance as
the price of survival.

April 24

I wonder if there's a way out of this . . .
if it's not too late to upend our current
internet model . . . to come up with new
vibe, a new ambience, a new way for
the people of the world to communicate
with each other. One that doesn't harvest
our personal data and sell it to the
highest bidder . . . doesn't feed on anger
and disinformation . . . doesn't cause
irreparable damage to our children, our
mental health, our democracy and our
cultural and emotional lives . . . doesn't
kill our spirit and turn us into manic
scrolling addicts.

> Hackers won't do it . . . grilling the
> mindlords in front of congressional
> hearings won't do it . . . the European
> Union's Open Internet protocol won't do
> it . . . the only thing that can clean up the
> toxic areas of our mental environment
> is the birth of a Mental Liberation Front
> (MLF) — a global movement of angry,
> mindfucked people acting out with feral
> ferocity and all-out civil disobedience
> against surveillance capitalism
> . . . hacking and crashing Big Tech's
> platforms . . . bringing the current system
> to its knees . . . and then, on its ashes,
> building a new one . . . a new way for
> us humans to talk to each other . . . one
> that does not mix communications with
> commercialism — a FREENET.

April 25

The MLF has an intensely personal side too . . . it's about gaining agency from the neck up . . . learning to nip anxieties

in the bud, modulating mood swings, shifting vibes, ambiences, tones, funks . . . seeing dark clouds on the horizon and taking action before they engulf us . . . and being fearless at just the right moment.

MLF Manifesto

Rise up, hybrid creature!

Break free!

Liberate yourself from algorithmic mindfuck!

Halt the secret extraction and manipulation of your personal data!

Destroy surveillance capitalism! Topple the mindlords. Demolish the spy panopticon.

Fight, fight, fight for a new Declaration of Human Rights . . . a new Bill of Rights for our kind.

Digital-body sovereignty now!

The Third Force

April 28

Last night a wonderful
high . . .
like in meditation or
yoga . . . a liberating
lightheadedness . . .
a surging optimism . . .

a mental glow.

bioconsciousness

You have to wonder how our current time of bumbling will be remembered. Typically, we slap a label on every era of human history, and usually it's about a monumental tool we invented or a cultural transformation that fundamentally altered our lives. The Bronze Age. The Renaissance. The Modern Age. The Space Age. The Digital Age. The Genocidal Age. But now it looks like our era will be known not for a leap forward but a slide back. Future generations will remember us for the damage we did: the carbon we spewed, the plastic we dumped, the forests we clearcut, the species we exterminated. And also for the financial meltdowns we unleashed, the inequality we tolerated, the surveillance we endured, and the epidemic of mood disorders, anxiety attacks and depression we inflicted upon ourselves.

Instead of the Anthropocene Age, why don't we call it what it really is: The dawn of a Dark Age, a time when we turned away from nature and each other and ran our planet to the ground.

Of course, we still have a chance to turn this story around. But it's going to take an attitude conversion — a new way to behave as a species.

The first seed of this new perspective was actually planted half a century ago.

On Christmas Eve, 1968, the first startling images of the blue Earth, taken by Apollo 8 astronaut Bill Anders, were beamed back from space. There it was, sparkling against the void, home field, the source of all known life. The deep and sudden urgency those pictures aroused dimmed a little as the information age sent us back into our collective neocortex. But we'd glimpsed our place in the order of things. Our planet as a biological organism in the cosmos and we humans scrabbling around on its skin with many other living things. It was a mind-blowing insight, and it's still there, buried under our fears and furies.

The task now is to rekindle that original spark, and take it to the next level.

Every one of the 200 countries on earth is itself a living organism — an autonomous component within the larger biological unit. Every citizen is a cell within the body politic. Very likely Gaia herself is one of many living planets peppered throughout the universe.

If that sounds overly touchy-feely, it's only because our horizons have been squeezed so radically by hyper rationalism that we can no longer feel the deeper dimensions of being human. But now that we're entering the planetary endgame, we'd damn well better find a way to feel it again.

If you understand your country as a fragile, hypersensitive, living organism, the questions become very basic:

How do you keep her fed and safe and self-sustaining? How do you keep the people happy? How do you encourage them to play fairly with each other? How do you empower them to make wise decisions?

Suddenly "politics" is no longer just a bloodless exercise that we can choose to dabble in or not. It isn't just about reading indicators (housing starts, polling integrity, polity scores) or tweaking the mixing board (electoral boundaries, reserve requirements, cabinet size). It isn't about crowding under the umbrella of your nation's military might or rallying around its sloganeering. It isn't even about being on the Left or the Right.

Instead, it's something else entirely, something personal and universal at once, an overarching concept that captures every aspect of being, staying alive and thriving on our fragile planet.

Call it *biopolitics*.

Successful societies of the future will gravitate towards the group-oriented Asian model: communal rituals will pull the people tight.

When emergencies arise people will come to each other's aid almost as a reflex. Life will be less about individual rights and more about shared values, traditions and destinies. As your circle of concern widens and rigid ego structures crumble, something becomes crystal clear, and you might even say it out loud: *My purpose here is to stay connected.* Disengaging isn't even an option. I'm part of this thing. I need it and it needs me.

And that is quite beautiful but also troubling. Because now pretty much all you can see is the assault on nature that's happening all around. The wounds we're inflicting. The poisoning and dumping and clearcutting and trawling and lopping off of mountain tops. The plastic microbeads choking the oceans.

It's all feeling very personal now. You're wondering about the role that you and your family and community and country are playing here. And the corporations. *Especially* the corporations. These legal entities that we ourselves created have grown into monoliths that grind the bounty of the natural world into income for shareholders. Suddenly you see them for what they really are: tumors in the flesh of the Earth. Pathogens that attack her immune system and will surely kill her unless We the People learn to control them.

And then you start thinking about the blood that flows through her circulatory system: the money.

Money plaques up dangerously all over as moneylenders accrue interest and money handlers charge fees. Convoluted financial instruments further slow the blood flow. Blood pressure goes through the roof. That is the menace of runaway financialization: it clogs up the arteries. We have to get rid of the middlemen and let the blood flow freely again.

The pumping heart of the system is the Federal Reserve, which was created a century ago to stabilize the banking system after a series of panics caused wild market spikes and runs on banks. But after the financial meltdown of 2008 the Fed went into serious tachycardiac overdrive, pumping trillions of dollars of "quantitative easing" into the system. This was like junk food to a sick patient. It gummed up the works even as it fed Big Finance, allowing it to grow ever bigger and more predatory. Now, as autonomous armies of bots make flash trades and markets pitch and swoon, a massive heart-attack could happen anytime.

And now that we're knee-deep in the metaphoric gore, you may be thinking a little differently about your role in all this. You're positioning yourself within systems within systems. From now on, you decide, I will live my life differently. I will support local business not just because it gives me personal satisfaction, but because everyone needs secure access to the stuff of life. We need thousands upon thousands of local farmers spread across the land if we're going to win the planetary endgame. Sure, supermarket veggies are cheaper and processed food even cheaper, but that kind of shopping is no longer part of my life philosophy.

If you tack the prefix "bio" in front of every facet of life, policy decisions and new ways of living snap into clarity.

When we draw up a "true cost" pricing scheme to restructure markets so they align with ecological sustainability, that's *bioeconomics*.

When we talk about replacing mass, for-profit incarceration with a restorative-justice model — a systemic solution that invests the whole community in the problem of crime — that's *biojustice*.

When we organize en masse through the flocking signal of social media, that's *biocommunication*.

When we grow our own food, that's *bioconsciousness*.

So it goes. Your economy is a *bioeconomy*, in the sense that every human action pushes on something that pushes on something that puts a dent in the natural world. Every market transaction is a *biotransaction* that impacts the earth one way or another.

At root, all politics is *biopolitics*. Maybe this is what Aristotle was talking about when he said, "man is by nature a political animal." At some level we ache to be in the churn of the collective struggle, performing some useful function within the whole.

The endgame is survival. Economists think "short run," and "long run," but there *is* no long run if you get these decisions wrong. The fundamental questions all of us should be asking become front-and-center:

Who owns the water?
Should nature have rights?
What voice should local communities have over resource extraction?
Should future generations have a say?

Job One, then, is to start re-writing our constitutions with such questions in mind. These documents are the embodiment of our ideals and aspirations, our beliefs and values. The constitution thus becomes a living thing, flexible and nimble — not something written in stone long ago and worshipped like holy writ. It becomes a map of a future that's constantly moving, being recalibrated, reinvented, fought over and born anew with every generation.

This is how the human-centered 20th century model of biopolitics eases into biocentric mode.

This is how a new vision, a new playbook, comes to be.

The impact of this mindshift is felt inside the chest of everyone. You wake up in the morning and think, "Hallelujah" my country is alive, and I am part of it. I'm going to nourish this organism: feed it, prune it, protect it. My involvement in politics is as natural as fish schooling up. I shed *a portion* of my self, my rights, my me-first instincts, to serve a greater good.

The personal payoff is that, as your politics expands and matures in this biocentric way, you feel less lonely, more involved. And a new personal mantra bubbles up.

I connect, therefore I am.

As Bruno Latour so poignantly pointed out before he died, bioconsciousness holds out the promise of a renaissance-like reset of the sciences, the arts, the law, and politics . . . it offers us hope for a new kind of civilization: one in which we put aside human mastery and learn the 'languages' of rivers, mountains, oil pipelines, baboons, voodoo dolls, and viruses. For these are the many murmurs of Earth itself, growing louder and louder.

I AM BECAUSE WE ARE

— African philosophy of Ubuntu

April 12, 2033

rituals . . . global rituals . . . celebrating everything from the Spring Equinox and the UN charter, to work, play, music, art, frugal living, fish, frogs . . . these are what bind us together now.

So now we pause to cast an eye across the landscape.

At this point you've seen most of "the codes": the first six metamemes that will pull us back from the brink. To recap:

- A true-cost marketplace (where the price of all goods and services tell the ecological truth)
- The end of secrecy (fostering a climate in which the people must know everything, because knowledge is power)
- A corporate charter revocation movement — to identify the most carceral companies and wipe them off the face of the earth
- Bionomics (economics with diversity, locality and contained growth as its pillars)
- New brakes on fast money
- A Mental Liberation Front (where we regain the ability to think for ourselves, feel deeply, and shake off crippling fear and depression and shame)

But none of it will happen without the seventh: a shift in the tone of the world. The vibe of life.

The world has changed almost beyond recognition in the past decade. A sudden psychological drama has unfolded in stages, as one curve ball after another came slamming in.

Climate change scared the hell out of us.

Then came Covid. And that forced social experiment revealed massive social problems that had been building under cover of the chaos of daily life.

Then the sudden new reveal about just how far along machine learning actually is. This Promethean new entity called ChatGTP. It's as if AI suddenly pulled open the dressing-room curtain to reveal where it is now, what it is now. It has claimed this space in our lives, hinted that it won't be long till it is basically running our affairs. And then came the AI powered slaughter in Gaza, along with an upping of the surveillance game, beyond all borders. Using facial recognition to stamp out all forms of resistance. And it has forced us to ask: at this point how much agency do we even have?

On top of it, widespread questioning of social, scientific and biological norms. Is there such a thing as a fixed gender? And what is that 85% of the universe called dark matter?

All of these forces — each of them seismic in its own way — more or less landing on us at once, have produced a historical moment that is, arguably, bigger than the Enlightenment. Bigger than Modernism. Bigger than any shift in the last 20,000 years.

Every era has a vibe, an ambience, a way of knowing and experiencing that permeates our existence.

It is the stamp of what it feels like to be alive, the pulse of the zeitgeist. It's an intangible thing, hard to pin down when you're in it. But its gravitational pull is so strong that it bends history.

The Beats gave us permission to be wild and carefree. The Situationists taught us to live without dead time. The hippies pushed us back into nature. The punks tuned our radar for hypocrisy and our will to resist. The foot soldiers of Occupy Wall Street made us believe that world revolution is possible — and set the tone for what's to come.

More recently, movements like #MeToo and #BlackLivesMatter have fundamentally changed the way we feel about the most intimate aspects of our lives, and what human relations are really about.

Our crisis is a crisis of aesthetics

Art and design movements — from Impressionism to DADA, De Stijl to Design Anarchy — however briefly, catch the essential spirit of their time. "Many of the great cultural shifts that prepare the way for political change are aesthetic," J.G. Ballard said, and of course he's right.

But now, at the beginning of the 21st century, something feels ominously different. We find ourselves in a planetary endgame, a "code-red emergency" — gripped by an unprecedented existential tension. This calls for wild, urgent, creative, hair-on-fire innovation. Instead, we have . . . straight-line thinking.

Last year, I happened to turn on the TV during the baseball playoffs. One team was facing elimination — lose this game and they were done. This team had its best pitcher on the mound, and he was pitching out of his mind. Unhittable. Untouchable. Damn near shamanic. The performance was giving everyone the shivers, no matter which team they were rooting for.

Then the manager walked out to the mound and pulled him off the game.

See, the quants in the front office had determined that when a pitch count hits 70, pitchers start to lose their stuff. On average. And so management hedged its bets. It busted in there and broke the spell and killed the magic that was materializing in front of their eyes.

They couldn't see it. They couldn't feel it.

In came a new pitcher. Who promptly gave up the tying and winning runs. And just like that it was over.

I thought: *This is why the world is imploding.* This is what straight-line thinking does: it destroys everything it touches. It has no clue of the damage it's doing.

Hyperrationality. The master narrative of life on earth. A holdover from the Enlightenment, with no real course correction in 2,000 years. It is the worst tool imaginable for the job of pulling off a global mindshift. We shuffle into our apocalyptic future with nothing but our buttoned-down executive brains. We speak in corporate jargon and techno-bites. Our whole lives crushed down to probabilities, data points, Bayesian calculations. Run the numbers. Protect your priors. Quant it out. You can't have enough pixels on your phone camera or RAM in the cloud.

We shape our tools, and afterwards our tools shape us. This is where we've ended up. The media engines we made have re-made us. We think in algorithms now — some of them sound, some of them unsound, but all of them running by themselves like headless chickens.

In that spirit we hatched the *First Things First Design Manifesto 2000*. Thirty years earlier, UK designer Ken Garland had speartipped a crusade basically accusing designers of selling out. Now we recreated his manifesto with a vengeance. We said to designers, Look: You are among the most powerful people in the world. You are to the information age what engineers were to the age of steam, what scientists were to the age of reason. You set the mood of the mental environment — the look and lure of print, the tone and pull of TV, the knack and smack of the Net. You are the very *form* of our culture . . . the typesetters of thought . . . the editors of sentiment. So for chrissake,

Stop kissing corporate ass!

wake up and start acting like it! Unprecedented environmental, social and cultural crises demand that you move away from product marketing towards the exploration and production of *a new kind of meaning*. Instead of using your skills and imagination to sell sneakers, detergents, hair gel, butt toners, credit cards, light beer and heavy-duty recreational vehicles — *instead of kissing corporate ass* — why not break out of the commercial design box and start playing around with the eco- and psycho-dimensions of the product-in-use. Inhibit impulse, scramble habits, modulate desire, nudge human behavior in brave new directions. Steer us away from the hubris and self-destructive chaos of the planetary endgame. Design things that raise goosebumps on people's skin. Become a design anarchist: Riot . . . Kiss . . . Love . . . then Design! Help create the new wobbly vibe that will save the world.

More than a thousand designers signed our manifesto, including 22 of the biggest names in that world. And as we talked to young designers in the years that followed, we could see their thinking scaling up.

We'll use recyclable materials.

Nice, but it's not enough. We'll work for nonprofits. Okay, but what's the vision?

We'll work together to co-create the mojo that stirs people's souls, disrupts their toxic routines and tips this whole human project toward justice.

Now you're talking.

This is the pivot our whole culture now needs to make.

The aesthetic of our time needs to be kicked out of the orbit of its destructive formalism, its whoring corporate supplication. Dirty it up. Summon the trickster out of the shadows. Scavenge value from the margins, from the stuff we reflexively throw away. Generate an era of tumult that wipes out every remnant of the old vibe.

"Order has failed so let chaos prevail!"

And so, the grand reset begins.

FROM MO TO POMO TO SLOMO

Modernism laid down the rules — "form follows function" — that programmed a whole century. It built planned cities, tone-deaf buildings, urban social housing that proved so soul-crushing it finally had to be leveled. Monster homes plopped on the seashore, boxes upon boxes, right to the property line, the curve of the shore against the diminished straight-line imagination of developers.

Postmodernism arrived as a correction — but it went too far, and the ground turned to liquid under our feet. Everything meant anything so it meant nothing.

Now we're on the cusp of a new era. We don't know what it's about, don't even have a name for it yet. *Slomo?* All we know is that we're in a winner-take-all race to find a new aesthetic, a new story to make sense of our code-red moment. We'll stumble towards a new sensuality, a new structure of feeling, a new sense of spiritual purpose . . . or we'll zombie-walk into a new dark age.

What *Slomo* won't be is the popular sci-fi vision of next-level techno-hyperrationalism; that is the wet dream of the logic freaks sliding into China's social credit model of algorithmic living. More likely Slomo will look to nature for its codes. Biomimicry, ribosomal thinking. Us humans insinuating ourselves into the web of life. It's like the difference between rowing and sailing. Rowing is brute straight-line effort that seems powerful and efficient, but you're never going to cross the ocean that way. The secret is to tap into the currents and the wind. The solo rower is doomed because as soon as you let go of the oars for even a second, you grind to a halt. But under sail, once you get yourself aligned with the forces of nature, you can lie back and dream.

In his recent book *The Day the World Stopped Shopping*, J.B. MacKinnon visited the Japanese island of Sado, where type-A Japanese, burned out from life in Tokyo, flee to recover their sanity. Here he heard the word "utori." It has no direct English translation but means something like running "below capacity," having a little slack in the line. Several people defined utori as having "space in your heart." I am available to you because I'm not overwhelmed by rat-race demands. I can give you my time and my attention and my energy and my thoughts. Yes, yes. Come on over.

MacKinnon had been around the world chasing the new post-consumerist emotional upgrade that will save us, and he may have found it in this one word. On some level, everyone he'd met who was living an intentionally small, happy, minimally destructive life was positively brimming with utori. You put people first. A spirit of soulful generosity becomes both the bedrock of your private experience and the face you present to the world. People with utori, MacKinnon concluded, "are simply better at being human."

As the temperature rises, screws tighten and the doomsday clock approaches midnight, all of us face a moment of reckoning, grappling with the most personal questions there are. Questions like: Would I rather be very rich or very spontaneous? Which person would my children rather be around? Who's more at peace? Who's living more gently on the planet?

People have had it with culture and politics in rational overdrive.

They're done with risk-management software, talking points on teleprompters and expertly concocted carbon-reduction plans that ultimately mean nothing. (That is, they sound good but don't actually make a dent in the problem.) We ache for proof that full-blooded living is still possible out there.

I think this explains why so many Americans voted for Trump.

A lot of people were willing to forgive the crazy-ass stuff coming out of his mouth because they were sick of the law-school elites running Washington. They were desperate for someone who shoots from the hip and doesn't give a damn about making a mistake or being cancelled. Trump walked up to a gaggle of journalists every day like a trumpeter ready to play jazz — with no firm idea of what was going to come out of his horn.

Miles Davis once said: "The biggest challenge in jazz improvisation is not to play all the notes you could play, but to wait, hesitate — to play what's not there."

Trump did that. And that level of spontaneity was, in his hands, a kind of black magic. That's what made his base so passionate and his rallies so raucous. It's how he was able to command incredible loyalty, to beat Kamalia and keep his grubby little fingers on the levers of power.

Every one of us needs to learn to live a little more like that. When we Lefties demonize Trump, we lose the lesson: If you're afraid of what the next note will be, you're not going to be able to play it. Pulling punches and playing defense all the time, like we've been doing for the past 20 years, that's not going to get us anywhere. Sooner or later you have to let it all hang out and go for it.

One shot, one life!

To return to the original question this book poses: What *is* this revolution I'm speaking about? A full-on Marxist insurrection? A dismantling of all structures of power?

Or is it something more personal and actually deeper for all of that? A revolution of everyday life?

Nothing is going to happen unless *you* change. Unless we all do.

The nut of this new enlightenment is action. It's about becoming an active change agent.

Change how?

By finding a new rhythm.

Go to the mirror, right now. Take off all your clothes. Stare at yourself for five minutes.

Stand there until the brain fog begins to lift.

The Zen poet Basho would say that this is partly an exercise in *muga,* or self-forgetting. You are forgetting who you used to be. What your culture demands you be.

At this moment you're not in any tribe. Not Left, not Right, not anything really. Just a human being caught naked in the flux of life. Ready for whatever comes next.

Wolfram Eilenberger's book *Time of the Magicians,* about Wittgenstein, Benjamin, Cassirer and Heidegger, had me mesmerized for a good part of the year. I felt that these four men who re-invented philosophy after WWI, were on the same life journey as I was . . . and it was wonderful to walk alongside them for a bit.

But halfway through the book I had an epiphany. It occurred to me that all four of them were firmly *stuck* inside their heads. Heidegger secluded in his Black Forest hut. Cassirer in his patron's vast library. Wittgenstein never able to break out of his anxious brooding.

In fact, many of the great Western philosophers — Descartes, Newton, Locke, Pascal, Spinoza, Kierkegaard, Leibnitz, Schopenhauer, and Nietsche, were all closet thinkers — they never formed intimate ties, or reared a family.

My god, I thought: could this kind of obsessive thinking — our default mode since Aristotle, Pythagoras and Plato over two thousand years ago and the one Descartes captured in the five words: *I Think Therefore I Am* — be the fatal flaw of Western civilization? Could this be where we veered off course . . . where we fell into the trap of *thinking* that we can *think* our way out of everything?

Now after finishing the book, I wonder if that kind of compulsive thinking may no longer be fit to solve the multiple crises that now confront us. The vibe that will save us is: *feel first.* Use "thinking" mode for checking and correcting. It's only by feeling that we can

a future without artifice

surprise ourselves, and it's only when we surprise ourselves, as Ray Bradbury put it, that we find out who we truly are.

Maybe the vibe of the 21st century will reveal itself when we take the leap from *I Think Therefore I Am* —> *I Connect Therefore I Am*

Scientists used to think there were only six human emotions — anger, surprise, disgust, enjoyment, fear and sadness. We now know there is a seventh: awe.

Awe is the feeling, registering more in our body than our mind, that we're in the presence of something so vast and deep and powerful that it swamps our present understanding of the world. A skyful of stars in the middle of nowhere. A soaring piece of art. An act of wild kindness, fugitively glimpsed.

Or an existential threat.

Awe is unlike anything else we humans experience. When it shoots through us, we enter a kind of altered state. We become superhuman, sleeper agents suddenly activated for a purpose we never prepared for, and barely understand.

Awe interrupts the trance, silences the static and replaces it with the clarity of mind to understand what our job is now.

The craving to be part of a bigger project, this is the "collective effervescence" Emile Durkheim spoke of: an energy and harmony that only happens when we huddle up on strong groups for a shared purpose.

People suddenly understand where they fit in the story of the world, now and going forward.

The way forward may well turn out to be a new mode of activism, one that's less about street fighting and more about spiritual insurrection.

Maybe we will simply let go . . . cultivate a certain looseness of mind . . . "*Live suddenly without thinking,*" as e. e. cummings put it . . . and fall into whatever comes next.

Which is what?

Nobody knows . . . but it's tantalizing to speculate.

Maybe we'll start reversing a lot of things . . . shed many of the constraints we never asked for and choose their opposite. The stuff of the real world over digital simulacra. Traditions over fads, mystery over certainty, pathos over logos, listening over pontificating, child's play over exegesis, the collective over the individual, yin over yang, sharing over consuming, the long-time horizons of the planet over quick payoffs and indulged cravings.

Maybe as the techno-rationalist Western vibe wilts at the task before it, the so-called WEIRD vibe of the rest-of-the-world — communitarianism, family, tradition, faith — will rise as an alternative.

Maybe the new vibe will be a maker vibe — more about creating (which makes you powerful) than about consuming (which gives away your power).

Maybe we'll opt for small rather than imposing things. Muted colors. Understatement over hyperbole. Not the action but that pregnant moment before the action.

Maybe we'll abandon the spectacle . . . and revel instead in the intimacies of everyday life: the touch of a lover, a chat with a bright-eyed stranger, a quiet moment in the wild.

Maybe the old American dream about prosperity will morph into a new dream about spontaneity.

Maybe we'll have mystical feelings of oneness with nature!

Maybe we'll learn to start having . . . fun again. Just crazy, uninhibited fun. (Where did *that* go?)

Only connect!

Whatever form it ultimately takes, *Slomo* is the project of our century. And in its emergence, just maybe, will be the answer to a question that has not yet been answered: Is capitalism with empathy even possible?

One thing is certain: If we don't get it right, if we cannot recover our innate empathy, find balance and come up with a new tone, a new ambience, a new aesthetic, a new sense of awe to live by, then it will be a century of hubris, brutality and mayhem on a scale the world has never seen.

April 29 2033

straight lines are morphing into curves . . . rectangles into circles . . . certainty into wishful musing . . . formulas into jazz . . .

~~Capitalism~~
~~Feminism~~
~~fascism~~
~~Environmentalism~~
~~postmodernism~~
~~Terrorism~~
~~Populism~~
~~FEMINISM~~
~~Populism~~

anarchism

May Day

Amazing how this anarchist credo which for centuries operated in the underground and was always seen as a somewhat grubby ideology, championed by subversives like Kropotkin, Bakunin, and Bookchin . . . that this wild, fuck-it-all-and-play-jazz way of thinking and living has turned out to be the grand philosophical vibe of our time.

June 7

There's a real healer among us! She's a one-of-a-kind old soul.   I don't know how we ever survived without someone like her before the crash. She applies herbs from the garden to my cuts and bruises and advises me on what I need to eat more of based on the tinge of my iris. She checks my pulse and tells me I need to laugh more and start eating meat again. A few days ago she looked at my tongue and told me I needed to make love to Lil daily at sunrise for one month. My old shrink never said stuff like that to me. And when Lil has bouts of anxiety . . . she not only calms her down but transports her to this whole new dimension of insight.   I twisted my ankle the other day and as she was fixing it . . . it was as if she was fixing every other part of me that's ever been twisted and mangled and broken.   Sure.

She teaches us how to heal our aches and pains and our digestion. Our ingrown toenails and our infections.   But what she's really doing is teaching us how to live and die.

sea levels will keep rising for the next 1000 years or more

July 16

I wonder if this dark age we're entering into isn't some kind of existential riddle, a great spiritual test we must live through before we can see the light?

<u>Is there a final secret that will eventually be revealed?</u>

Sept. 17

my heart spikes with mad joy
at the sight of a patch of wildflowers

epilogue

I am 83 years old. Sometimes I feel like simply saying, as Wittgenstein did, "Tell them I've had a wonderful life," and call it a day.

But there's one thing that continues to excite, tantalize and bedevil me. It is a vision of what it means to be a political animal in the 21st century.

Somewhere, rooted in an ancient philosophy brought forward into contemporary times, lies the way out of this mess we're in. It is a politics beyond politics, a system that will bury the whole Left vs. Right paradigm for good.

What's emerging this time is not a movement of political parties or of governments; it's a movement of regular people, coalescing in cyberspace and pulling off big idea, metamemetic insurrections in the real world. It behaves uniquely and evolves organically. Within its many-chambered heart lies a secret so fucking dangerous and beautiful that it opens up all possibilities for governing ourselves in the future.

This is the *Third Force* — a bottom-up movement with the power to upend 20,000 years of top-down rule.

If we brainstorm globally, learn to act as one, and push ahead on the Eco, Psycho, Corpo, Econo, Politico and Aesthetico fronts, then we can work our way out of the existential fix we're in and begin a new chapter in the human story.

One thing to keep in mind: Revolutions don't begin with thoughts. They begin with "a mood, with astonishment, fear, outrage, worry, curiosity, jubilation." A critical mass of people gets dragged into the fight by a delicate alchemy of feeling . . . plus an element of pure luck.

The avocado falls from the tree when it falls.

Occupy Wall Street took off because the moment was ripe. Arab Spring had just happened. Mubarak was deposed. Greedy bankers had crashed the economy. Young people couldn't find jobs. The 99 percent was furious at the 1 percent. And along came a lightning rod for that anger. OWS galvanized around Zuccotti Park but fed off proxy energy from around the world.

That's when the *Third Force* was born.

Now, an even more virulent alchemy of conditions is in place. The screws are tightening on humanity. The future no longer computes. A massive destabilizing discontent is brewing. All the conditions for a world revolution are in place.

poof

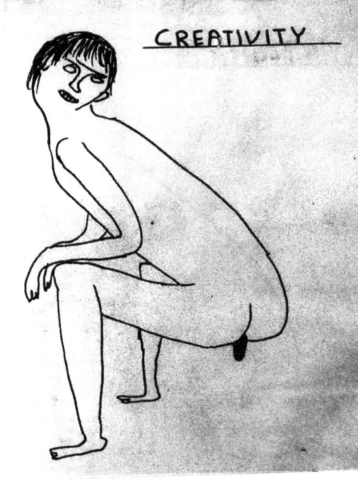

Graphic Credits

Creative Director: Pedro Inoue
Art Director: James Callaghan

New Ways To Live, Love and Think

Adbusters Logo, Adbusters
Occupy Wall Street poster, Kalle Lasn, Pedro Inoue & Will Brown

Birth of a Politics Beyond Left and Right

10 Years to Save the World, Pedro Inoue
The Third Force, Kalle Lasn
Meme Pyramid, Kalle Lasn
Meme Tree, Kalle Lasn
Finger pointing, Adbusters
Play Jazz photo, Jean-Philippe Ksiazek/AFP

One Shot One Life

Painted line, Adbusters art department

Culture Jamming — Let's Have Some Fun

Fuck this shitty distopia, Steve Vogel
New York Culture Jam, Adbusters
Man in a Can, Ethic-Ease, Adbusters
Anti-smoking anti-ad, Terry Shukle, 1976
I miss my lung, Bob, California Department of Health Services, 1998
Absolut Impotence, Adbusters spoof ad, Svea Reklam
Reader-submitted spoofs, various artists
Obsession for Men, Nancy Bleck
Obsession for Women, Thomas Antel
Shoe, Concept: Maquila Solidarity Network; Photo: Chris Gergley
Sweatshop Nike, Adbusters
Escape the Fantasy, Gary Baseman, garybaseman.com
Look Honey, Adbusters
Buy Nothing concept: Ted Dave
Escape Capitalism barcode, Marc Dejong
Carbon Dioxide, Adbusters
EXHAUSTED, Adbusters

The True Cost Revolution

The Bee Party, Adbusters
It's Pretty Amazing, Max Temkin
Ultimatum Poster, Adbusters
True Cost Bee, Adbusters
World Wide General Strike Poster, Adbusters, Photography by Stephen Shames
Bike Kill, Julie Glassberg julieglassberg.com

Battle for the Soul of Economics

Burning Bull, Adbusters

Bad Karma

Corporate Flag, Adbusters

Corporate Crackdown

Organized Crime, Adbusters
Bike Kill, Julie Glassberg, julieglassberg.com

Why Can't We the People Know Everything?

Constitution Amendment, Adbusters

A Fundamental Shift in the Nature of Value

Stock Market, Photographer unknown

The Creative Destruction of Neoclassical Economics

Cognitive Dissonance, Adbusters
A provocation projected onto the American Economic Association conference building, 2015. Kyle Depew.
Provocation at UBC, photo by Ruth Skinner
Copernican Shift, Adbusters
Suburbia, Brent Humphreys
Children playing, Rafiqur Rahman/Reuters
Kick it Over Manifesto, Adbusters
Still Alive, Artist unknown

Occupy Finance

Occupy poster, Pedro Inoue/ Midjourney
ATM Jam, photo by Joey Malbon; model: Ellina Rabbat
Great Pot Bang, Adbusters
The Bubble Burst, Midia Ninja

The Disillusionment

Blue Marble, NASA
Dog Walk, Mohsen Mahbob mohsen.carbonmade.com
Mask, Adbusters
I'm Not Real, Artist unknown
MLF Skater, Montage over photo by Ian Logan ianloganphoto.com
Moon, NASA

Bioconsciousness

Enso, Adbusters
Kowloon, Miya Moto

Vibe Shift

Burners, Thai Neave
Riot Kiss Love Then Design, Pedro Inoue
Well John, Ellen Lee
Crossover List, Adbusters
Days To Future, Holly Francis, behance.net/hollyfrancis
edge of the world, J.R. Hughto, flickr.com/photos/paintedland/
Dirty Fingers Girl, Photographer unknown

Epilogue

Global Uprising, Adbusters

Creativity, David Shrigley

Adbusters Media Foundation 1989 – 2023

DOUG THOMPKINS

CHRIS DIXON

PAUL SHOEBRIDGE

BRUCE GRIERSON

JAMES MCKINNON

MIKE (GOGGLES) SIMONS

JONATHAN BARNBROOK

MICAH WHITE

LAUREN BERKOVITCH

JULIAN KILLIAN

SARAH NARDI

DEBORAH CAMPBELL

HEI LAM NG

JISOO IM

JOEY MALBON

EDERSON MARTINS

TREVOR CLARK

WILL DAVIES

ELLINA RABBAT

JAMES CALLAGHAN

CHIARA MILFORD

MARCO VAN DER MEER

Unbreaking is a new imprint from 5m Books focusing on the regenerative action we need to take in response to the varied and interlinked challenges we face, encompassing agroecological, socio-political, economic and financial regeneration.

Unbreaking launched in 2024 with the publication of *Six Inches of Soil*.

Unbreaking seeks new books exploring the:

Agri-food system
Agroecological and regenerative farming
Alternative and local currencies
Cooperative and worker-led business structures
Degrowth
Ecological and alternative economics/funding
Land access
Social enterprises
Universal basic incomes/services

We want to hear from everyone about anything that will contribute to unbreaking our planet and our systems.